Lab and Activities Supplement
for

Statistics:
Concepts and Controversies
Sixth Edition

Dennis K. Pearl
Roger D. Woodard

W. H. Freeman and Company
New York

Printed in the United States of America

ISBN: 0-7167-2851-6

First Printing 2005

CONTENTS

Preface

Students learn in different ways. Thus, even sound "fixed-menu" teaching strategies will fail for some students. The philosophy of this Lab and Activities Supplement is for a "buffet" strategy offering students an assortment of ways to learn by supplementing their lectures and textbook with discovery laboratories, extra problems for review, applet activities, and problems from the Electronic Encyclopedia of Statistics Examples and Exercises (EESEE). There is a mixture of hands-on, software driven, and web-based activities as well as a mixture of individual and group/team activities.

But the collection is very incomplete. More web-based activities, EESEE activities, review exercises, and project investigations are being developed. To keep up with this fluid collection, check out the *Statistics: Concepts and Controversies* supplements Web page at www.whfreeman.com/scc. More exercises, copies of the data sets used in this Lab and Activities Supplement, and the CrunchIt software package can already be found there. The access code that comes with this manual will give you a full year's privileges on the site.

This supplement was developed for courses in statistical concepts that encourage students to think about the background of a problem, the assumptions of an analysis, and whether the goals of a study are being met.

Many helpful comments from lecturers, teaching assistants, and students have greatly improved the presentation in this supplement. The remaining problems are our fault. Let us know about them and the next version will be better.

Dennis K. Pearl
The Ohio State University
Pearl.1@osu.edu

Roger Woodard
North Carolina State University
Woodard@stat.ncsu.edu

LAB 1
Statistical Scavenger Hunt

Research reports, opinion poll results, government surveys, and other numerical presentations are a staple of life presented to teach us, to sway us, and occasionally to trick us. A goal of this course is to give you the ability to understand the statistical issues in the speeches you hear, the newspaper or magazine articles you read, the televised reports you watch, and the Web pages you visit. In this activity we will explore how the news media presents opinion polls and surveys.

Ola Babcock Miller was active in the late 19th-century women's suffrage movement, was elected to three terms as Iowa's Secretary of State, and in her first term founded the Iowa State Highway Patrol. Her election in 1932 surprised the political pundits of the day as she became the first woman, and the first Democrat since the Civil War, to hold statewide political office in Iowa. However, her election did not surprise her son-in-law, George Gallup, who predicted her victory using the first scientifically sampled election poll ever. Three years later Gallup founded the American Institute of Public Opinion in Princeton, New Jersey, and began publishing the results of the Gallup Poll in his syndicated column *America Speaks*.

At the Web site for The Gallup Organization (www.gallup.com) you will see how the Gallup Poll now provides data on all aspects of the attitudes and lifestyles of people around the world and how the poll is just part of a global management consulting and market research company. Gallup is now just one of many organizations who try to determine what the public is thinking by conducting scientific polls. In this laboratory you will explore how opinion polls are presented and what type of information is being left out of the stories.

FIND YOUR OWN
Search newspaper, magazine, or internet sources (other than gallup.com) to find reports of an opinion poll. Try to find a report that lists as much information as possible.

1. What is the title and source of the report you have found?

2. For the report you found of a poll, answer all of the questions below that are addressed in your article. The report you find must provide answers to at least five of these eight questions.

    a. Who carried out the survey?

    b. What was the population? What in the article indicated this as the population of interest?

    c. How was the sample selected?

    d. How large was the sample?

    e. What was the response rate?

    f. How were the subjects contacted?

    g. When was the survey conducted?

    h. What were the exact questions asked?

3. Carefully describe the parameter of interest in the poll you found. (Note: There may be more than one.)

4. Did any of the questions above help you to identify a source of bias in the conclusions of the report? Explain.

5. Do any of the questions above that were not addressed in the article give you concern about other potential sources of bias in the conclusions of the report? Explain.

6. Does the article present a margin of error that is associated with the poll? If so, what is that margin of error? Does it present the sample size?

7. Your instructor will ask each individual to write the margin of error from their poll on the board. Record the margin of errors and sample size for each student below.

8. Examine the relationship between the sample size and the margin of error. As the sample size increases, does the margin of error increase or decrease?

Does drinking milk help strengthen your bones? Does eating fish improve your thought process? Does a piercing in your tongue increase the chance of having receding gums? Each day we see features in the news media that indicate how aspects of our lifestyle might effect our health or welfare. In this activity we will examine research reports found by students in the class.

FIND YOUR OWN

Search newspaper, magazine, or internet sources to find news articles that present the results of research reports. Health information sites such as www.webmd.com or www.healthscout.com are good sources for these reports. This report should examine the relationship between two variables. Try to find a report that lists as much information as possible.

1. What is the title of the article you found? What is the source of your article?

2. What is the topic of the article?

3. What is the explanatory variable in this study? What are the different levels of this explanatory variable?

4. What is the response variable? What are the different levels of this variable?

5. Does the article detail an observational study or an experiment? Explain how you know.

6. Who are the subjects of the study?

7. What are some other variables that you think might influence the response variables other than the explanatory variable?

8. How does the study control for the variables you listed in the previous questions?

9. Was blinding used in this research article? Was it double blinded? Explain.

10. Does this study claim the results are "statistically significant"?

11. Does the article claim there is a cause-and-effect relationship between the variables? If so, are you convinced? Explain.

An observational study requires the casual observation of events, whereas an experiment implies active intervention on the part of the researcher. In this lab we consider what makes a good designed experiment.

CRITIQUING THE MEDIA
Consider the following news article.

Does TV make kids want to buy toys?

Watching television has been linked to many problems in children, including obesity. Could it also be linked to asking for toys? How do we make children stop asking for toys? Could it be just turning off the television? A recent study taught 88 third- and fourth-graders a series of 18 mini-lessons in school over a six-month period, aimed at helping them learn to watch television within a budgeted amount of time. The researchers, lead by Dr. Thomas Robinson, then asked them to limit their TV watching to just seven hours per week. Not all of the children were able to stay within this budget, but most reduced the amount of TV watched by about one-third.

To allow them to make a comparison, the researchers selected a control group of 87 students from a different elementary school. This control group did not get the training sessions and were not asked to budget their TV time. The group that watched a reduced amount of television were much less likely to ask for toys they had seen advertised on television.

1. What is the explanatory variable in this study? What is the response variable?

2. Does this study provide convincing evidence that reducing amount of television watched will influence children's requests for toys? What do you believe is the key weakness of this study?

DESIGNING YOUR OWN

We would like to determine if a reduction in the amount of television watched by children will influence the children's requests for toys. We have found 175 children from a local elementary school whose parents will allow them to participate in the study. Explain carefully how you would redesign the experiment that was described in the previous news article by answering the following questions.

3. Into how many groups would you divide the children? What are these groups and what differences would you create between them?

4. What procedure would you use to divide these children into the groups? Explain.

5. What other precautions would you use to insure that lurking variables do not confound your experiment? Explain.

LAB 4
What is Statistical Significance?

What is statistical significance? If a difference is statistically significant then it is large enough that we would not expect it to happen just by chance. If we randomly assign subjects to two groups we would expect there to be some difference in the groups just by chance. In this activity we will explore what size difference we should expect to see.

A college professor would like to determine if exam scores will be different if students take an exam on colored paper versus on white paper. She randomly assigns her students to the two colors of the exam. In this activity we will examine the differences that could happen by chance. We will do this by assuming that the color of paper does not affect the student's score.

1. Consider the following list of exam scores for a group of 20 students. For each of these scores flip a coin. If the coin lands heads up, assign the subject to the colored paper group. If it lands tails, assign the subject to the white paper group. Continue flipping coins until all of the subjects are assigned.

Score	Group
95	
90	
89	
85	
81	
81	
80	
79	
79	
78	
77	
75	
73	
71	
65	
61	
60	
59	
55	
50	

2. Record the average score for the subjects in each of the two groups. Also calculate the difference between these group averages (white paper average minus colored paper average). Be sure to indicate if your difference is negative or positive.

White paper: _____   Colored paper: _____   Difference: _____

3. Record your averages on the board where your instructor indicates to do so. Examine the averages reported by your classmates. What is the biggest difference that you observe? What is the typical difference that you observe?

4. Suppose the college professor found that there was a difference of 3 points in the average score of the two groups in her experiment. Do you feel this difference is likely to happen just by chance? Explain your reasoning.

5. Suppose the college professor found that there was a difference of 10 points in the average score of the two groups in her experiment. Do you feel this difference is likely to happen just by chance? Explain your reasoning.

LAB 5
Measurement Errors

In this lab we are going to collect some simple measurement data and use it to study reliability and bias in making measurements. The lessons of the lab are very widely applicable because essentially every piece of data can be viewed as the result of making a measurement. For example, a person's response to a political poll is a measurement of that person's beliefs; just as the value that the doctor records in a patient's record after taking the patient's blood pressure is a measurement.

WARM UP

"MERMAID" is a system for monitoring pollution in coastal waters, estuaries, rivers, and lakes developed by the GKSS Research Center in Geesthact, Germany (MERMAID stands for Marine Environmental Remote-controlled Measuring And Integrated Detection – see w3g.gkss.de/mermaid). Of course, the concept of "pollution" is very complex, so the MERMAID system measures dozens of different variables such as chlorophyll levels, water pH, ammonia levels, nitrate and nitrite levels, phosphate levels, silicates, water temperature, oxygen levels, etc. ... A MERMAID system is actually a collection of modules that transmits signals to a land-based station from an ocean platform, a buoy, or even an unmanned ship (the picture shows a MERMAID platform in the Wadden Sea). The GKSS Research Center is very particular about the accuracy of the devices they include in the MERMAID system.

For example, when the phosphate measuring module was used many times on the same water sample, the results were so consistent that they varied from each other by only a few parts per billion. Also, they compared their automatic measuring devices with the most modern "gold standard" method of conducting each of the chemical analyses and found that the results were essentially identical.

1. From the above paragraphs, explain in your own words how the GKSS Research Center is addressing the issues of measurement reliability, bias, and validity in their MERMAID systems.

MAKE YOUR OWN MEASUREMENTS

In this lab, you will use the "ruler" provided at the bottom of this page to measure the length of your statistics textbook. Team with a group of other students in your lab and decide on how you will make your measurements.

Cut along the dotted line to remove the "ruler." Write down your measurement to the nearest tenth of a unit (e.g., 6.3 or 12.7). Be sure that you don't influence your group members by telling them the value that you found or what measurement anyone else obtained until all of the measures have been made.

Record all of the values below. Your instructor will ask you to write these values on the board. Be prepared to describe in lab how the students in your group made their measurements. Also be prepared to discuss what possible sources of bias you saw in the procedure.

Our measurements _____ _____ _____ _____ _____

0	1	2	3	4	5	6	7

If the Pharoah's architect had used the ruler from Statistics class . . . .

IN THE LAB

Measurements are variable – they will come out a bit differently each time they are made. How different they are is seen in the *reliability* of the measuring instrument. Also, sometimes the procedure used makes systematic errors that affect the outcome in the same direction no matter how many times the measurement is made. These systematic errors are reflected in the *bias* of the measuring process. We are going to examine the reliability and possible bias of the measurements obtained by the students who measured the textbook using the crude "ruler."

2. Record the measurements made by the rest of your class. Carefully keep all the measurements from each group together.

3. Are there any groups that have very different measurements than the rest of the class? Could these have occurred from using a different procedure in carrying out the measurements? Explain.

4. What do you think about the reliability of measurements your class made? Is there a large amount of variability? Are the measurements of some groups more reliable than others? Explain.

5. If every measurement in the class were made by lining the ruler up with the side of the page rather than with the 0 mark, would this affect the reliability of the measurements? Would it affect the bias of the measurements?

6. The length of the textbook has been measured using an architect's ruler, and the value will be given to you by your instructor. How close are the individual measurements obtained by your classmates to this value? Are any of the values clear outliers? How close is the average of the measurements to the value given by your instructor? Was the average closer to this value when the outliers were removed? Comment.

7. Are the measurements made using the crude ruler *valid*? Explain.

Breakfast cereals are a staple in the diets of college students. But are cereals a nutritious alternative? In this lab we will explore the nutritional content of breakfast cereals. Most of these characteristics are numerical in nature, such as the number of calories or amount of sodium in each serving. To examine the distribution of numeric variables like these, we will make use of histograms. Throughout the lab we should remember that graphics tell us about the real world data.

AT THE COMPUTER
Use statistical software to open the data file called "Cereals," which provides the nutritional information for 77 brands of breakfast cereal.

Many diets in the first part of this decade espoused the idea that lower carbohydrates would be a way to lose weight. How many carbohydrates are in cereals? The variable "carbo" contains the number of grams of complex carbohydrates that are in each type of cereal.

> **Software Tip: Creating Histograms**
>
> In Data Desk® click the variable and choose **Histograms** under the **Plot** menu.
>
> **CrunchIt!** In CrunchIt ® choose **Histogram** under the **Graphics** menu, select the variable from the list in the dialogue box. Crunchit can be accessed at http:bcs.whfreeman.com/crunchit/scc

1. Using the data, find the cereal "Cheerios." How many complex carbohydrates does it have per serving?

2. Next, create a histogram of the "carbo" variable. Sketch this histogram here and describe its overall shape.

Examine this histogram carefully. Notice that across the horizontal axis are the amounts of complex carbohydrates and that the class intervals provided by your software are all equal in width. The vertical axis uses a frequency scale that indicates a count of the number of values falling into each class interval.

3. How wide are these intervals?

4. How many cereals have over 20 milligrams of complex carbohydrates?

5. Software packages like Data Desk and CrunchIt will allow you to identify individuals by selecting a portion of a histogram. Select the bar of the histogram that corresponds to the largest amounts of complex carbohydrates, and determine which cereal has the highest amount of carbohydrates?

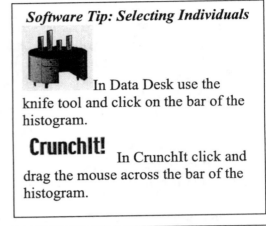
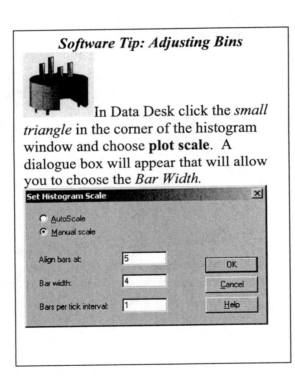
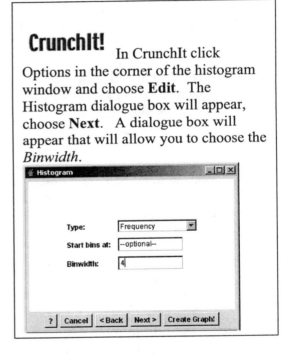
6. Histograms group numeric values into categories. If too few categories (also called bins) are used, some characteristics of the distribution can be lost. You can adjust the number of bins to see how this might happen. Adjust the width of the bars in your histogram to 4 milligrams. Sketch this histogram

7. Adjust the width of the bars in your histogram to 2 milligrams. Sketch this histogram. What key feature is visible here that was not visible in the previous histogram?

Often we want to examine several graphical displays and see the connections between the variables. Software such as Data Desk or CrunchIt allow you to see these connections. We will use this to see the connection between different manufacturers and the amount of carbohydrates. Select the variable that gives the company that manufactured the cereal and create a pie chart of these companies. Place this pie chart and your histogram of carbohydrates side by side. Click on the slice of the pie that corresponds to Kellogg's. As an example we have done this for *sodium* in the figure below. Notice that part of the sodium histogram is now highlighted as shown in the figure below. The highlighted part of the histogram shows the sodium content of the cereals made by Kellogg's.

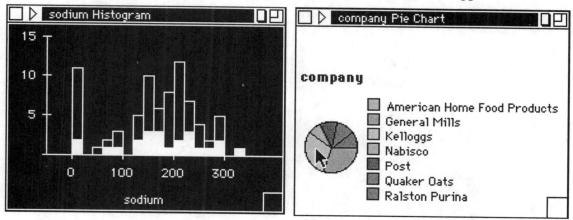

8. Approximately, what is the range (lowest to highest) of the carbohydrate content of the cereals made by Ralston Purina in this data set? (Base your approximation on whatever histogram you were left with after question 7.)

9. Which manufacturer sells the cereal with the highest carbohydrate content?

10. Which manufacturers offer cereals that are lowest in carbohydrate content?

11. How do the different manufacturers contribute to the overall shape of the histogram of carbohydrate content that you described in question 7? Comment about how each manufacturer contributes to the low, middle, and high areas of the carbohydrate histogram.

Rick Miller '92

In Statistics we look at both the location and the spread.

Sometimes the salient features of a histogram can be summarized by a few statistics such as the mean (or average), the median, the standard deviation, or the five-number summary. These quantities allow us to make comparisons of groups and help us to understand the story behind the data.

AT THE COMPUTER

Let's examine how these quantities relate to the histogram that they summarize using data that was gathered in September of 1999 when students responded to a survey in a large statistics class. Use your software to open the data file called 'Au99survey'.

1. Three of the variables in the survey were birthday month (1 for January, 2 for February, etc.), the amount spent on textbooks (in dollars), and the number of sodas consumed over the previous week. Without looking at histograms of the data, guess which of the three rough sketches below will approximately describe the histograms of the three variables. Explain your guesses.

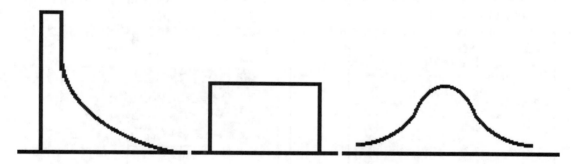

2. Create the histograms for each of these three variables. How do the histograms compare with what you expected?

3. Look at the three histograms for birthday month, textbook costs, and sodas consumed. Guess the value of the median, the mean, and the standard deviation for each variable.

Variable	My guess at the median	My guess at the mean	My guess at the standard deviation
month			
textbooks			
sodas			

To see how well you did at judging the descriptive statistics in question 3, use your software to compute the actual values.

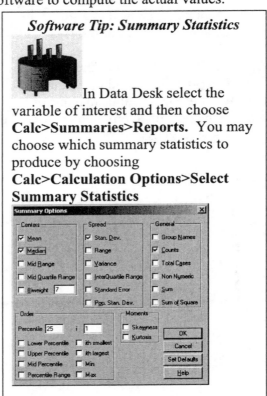

Software Tip: Summary Statistics

In Data Desk select the variable of interest and then choose **Calc>Summaries>Reports.** You may choose which summary statistics to produce by choosing **Calc>Calculation Options>Select Summary Statistics**

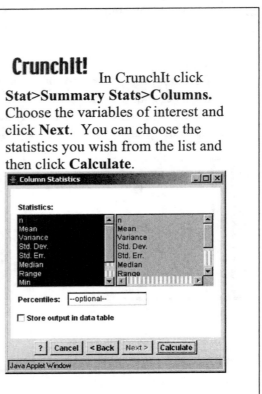

CrunchIt! In CrunchIt click **Stat>Summary Stats>Columns.** Choose the variables of interest and click **Next.** You can choose the statistics you wish from the list and then click **Calculate.**

4. What are the actual values for the median, mean, and standard deviation of the three variables?

Variable	Actual median	Actual mean	Actual standard deviation
month			
textbooks			
sodas			

5. Compare the actual values with your guesses. Which variable's summary statistics were most difficult to judge? Explain.

6. Which measure of location (mean or median) would be the most appropriate in each of the following situations? Explain.
   a. You want to know how many sodas a typical student in the class drinks per week.

   b. You want to know how much soda is consumed in a week by the class as a whole.

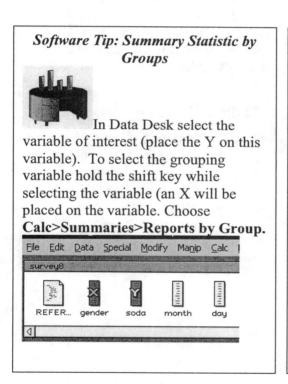

Software Tip: Summary Statistic by Groups

In Data Desk select the variable of interest (place the Y on this variable). To select the grouping variable hold the shift key while selecting the variable (an X will be placed on the variable. Choose **Calc>Summaries>Reports by Group.**

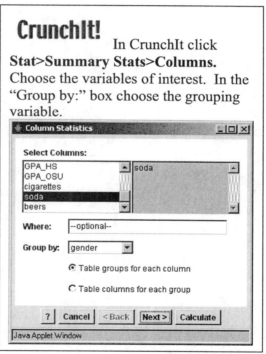

CrunchIt!

In CrunchIt click **Stat>Summary Stats>Columns.** Choose the variables of interest. In the "Group by:" box choose the grouping variable.

7. Do males drink more soda than females? One of the big advantages of summary statistics is that they allow us to compare groups such as males and females. Record the following summary statistics for the number of soda's consumed by males and females.

Gender	Mean	Median	1st Quartile	3rd Quartile	Min	Max
Males						
Females						

8. Based on your summary statistics, do you feel males or females drink more soda?

9. Boxplots are useful for making a graphical comparison between groups in a population. Use Boxplots to compare the High School GPA for the men and women in the class. Sketch the comparative Boxplots below.

> **Software Tip: Comparative Boxplots**
>
> In Data Desk select the variable of interest (place the Y on this variable). To select the grouping variable hold the shift key while selecting the variable (an X will be placed on the variable. Choose **Plot>Boxplot y by x.**
>
> **CrunchIt!** In CrunchIt click **Graphics>Boxplot.** Choose the variables of interest. In the "Group by:" box choose the grouping variable. Select **Next** and click **Use fences to identify outliers**.

Lab 8
Exploring Correlation

The correlation coefficient measures how tightly the points on a scatterplot cluster around a line. In this lab we will examine scatterplots and correlation coefficients for many pairs of variables. We will look at data from the EPA evaluation of 1999 model year cars, from a Statistics class survey taken in Autumn 1999, and from a class experiment that studied the relationship between forearm length and height.

AT THE COMPUTER

In this lab, we begin by gaining some practice at judging the values of correlations by looking at the scatterplots. We will begin by looking at a data set that deals with cars and gas mileage. The data set 'Cars99' lists characteristics of 140 car models from the 1999 model year. These values give information on items such as weight of the vehicle, highway gas mileage, and engine size. Correlations and scatterplots can help us understand relationships between variables for these cars.

1. With the recent increases in fuel cost many people are concerned with fuel mileage. We begin by examining the weight of the car and the highway fuel mileage. Remember that in a positive correlation, as one variable increases the other also increases. In a negative correlation, as one variable increases the other decreases. Do you think that the weight of the car and the fuel mileage (miles per gallon) would be positively related, negatively related, or near zero?

   Negative   **Near zero**   **Positive**

2. Engine displacement indicates the size of a vehicle's engine. In general, large or high performance vehicles have larger engines. What type of relationship do you feel this variable would have with fuel mileage?

   Negative   **Near zero**   **Positive**

3. Next let's consider the relationship between city gas mileage and highway gas mileage. What type of relationship do you feel these variables would have?

   Negative   **Near zero**   **Positive**

4. Finally consider the relationship between the weight of the car and the displacement. What type of relationship do you feel these variables would have?

   Negative   **Near zero**   **Positive**

<table>
<tr>
<td>

Software Tip: Creating Scatterplots

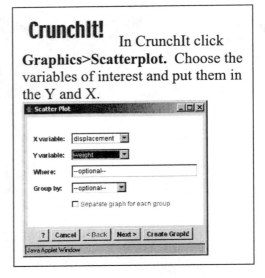

In Data Desk select response variable of interest (place the Y on this variable). To select the independent variable, hold the shift key while selecting the variable (an X will be placed on the variable). Then choose **Scatterplot** under the **Plot** menu.

</td>
<td>

CrunchIt! In CrunchIt click **Graphics>Scatterplot.** Choose the variables of interest and put them in the Y and X.

Scatter Plot		_ □ ×
X variable:	displacement ▼	
Y variable:	weight ▼	
Where:	--optional--	
Group by:	--optional-- ▼	
	☐ Separate graph for each group	
?	Cancel | < Back | Next > | Create Graph!	
Java Applet Window		

</td>
</tr>
</table>

Now let's see what the actual data indicates for these variables by making scatterplots of each pair of variables. Open the 'Cars99' data file and make scatterplots of the pairs of variables we previously discussed.

5. How did your predictions compare with the actual scatterplots? Did you predict any positive correlations to be negative or vise versa? Mention any differences here.

6. Examine the scatterplots you have created.
    a. Which of the correlations appears to be the strongest? Remember that a strong correlation is one that is tightly packed near a straight line.

    b. By looking at the scatterplots, what correlation would you expect for these variables? Make a guess rounded to one decimal place along with a direction (positive or negative). Write your guess in the appropriate space below.

Variables	My guess at the correlation	Actual correlation
'displacement' and 'weight'		0.834
'displacement' and 'mpg:city'		
'mpg:city' and 'weight'		
'mpg:city' and 'mpg:highway'		

c. Now calculate the actual correlation using software, and record those correlations in the table provided. Which of your guesses was off by the most?

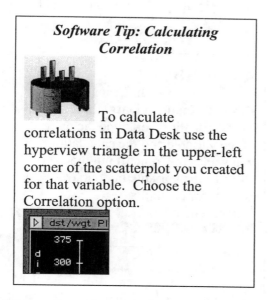

Software Tip: Calculating Correlation

To calculate correlations in Data Desk use the hyperview triangle in the upper-left corner of the scatterplot you created for that variable. Choose the Correlation option.

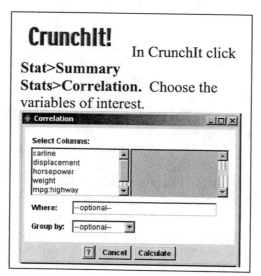

CrunchIt! In CrunchIt click **Stat>Summary Stats>Correlation.** Choose the variables of interest.

It is good practice to first take a look at the scatter plot before calculating the correlation coefficient in order to see if it is an appropriate measure of the strength of the association. For example, you should look for evidence about whether the pattern of association between the two variables is linear.

7. Does there seem to be a nonlinear relationship between any of the pairs of variables you examined? Which ones?

How does changing the unit of measurement change the correlation between variables? We can explore this by examining the conversion of the engine displacement and the weight of the car. Back in 1999 engine displacement was typically given in cubic inches, but now most cars list their engine displacement in liters. How do the correlations change when we convert cubic inches to liters? We can find out by calculating a new variable that divides engine size by 61 (there are approximately 61 cubic inches in a liter).

Software Tip: Creating a New Variable

To create a new variable click **Manip>Transform>New Derived Variable.** Give the variable a name of your choice and click **OK**. A window will appear in which you should type the formula for the new variable. In Data Desk be sure to put the variable name in single quotes. For example: 'displacement'/61.

CrunchIt! In CrunchIt click **Data>Transform Data.** In the Transformation box, type the name of the variable and the calculation you want. In CrunchIt put the name of the variable in double quotes, for example: "displacement"/61.

8. Create a scatterplot of your new variable and the weight of the variable. Examine this scatterplot and the scatterplot of displacement and weight you made earlier. How does the pattern of this scatterplot compare with the previous scatterplot of these variables?

9. Calculate the correlation between these variables. How does this compare with the correlation you found between these variables previously? Explain.

Now let's switch to another data set that deals with students and a survey they completed. Open the data file called "Au99survey," which gives the results of a survey given to 401 students who took a Statistics course at Ohio State University in September 1999. We'll look at the variables 'height' (the students' height in inches), 'mate height' (the height of the students' "ideal mate"), 'year' (of birth), 'age' (in years), 'HS GPA' (high school grade point average on a four-point scale), and 'OSU GPA' (grade point average at Ohio State).

10. Before looking at the data, make a guess at the size of the correlation for the pairs of variables listed below. Record your guess in the table below. Next make a plot of each pair of variables from the "Au99survey" data file. Look at each plot and try to guess the value of the correlation. Record your guess in the table. Finally, use the software to find the actual value of the correlation between each pair of variables and record that value.

Variables	My guess at the correlation	My guess after looking at the plot	Actual correlation
'height' and 'mate height'			
'year' and 'age'			
'HS GPA' and 'OSU GPA'			

11. Which of the three correlations in the previous question were the most difficult for you to guess? How did the three correlations differ from your expectations with respect to direction and/or strength?

12. Many people are surprised at the direction of the correlation between the students' height and the height of their ideal mate in this survey. Think of an explanation for this paradox and use the software to investigate your explanation. Show (sketch or cut-and-paste) the results below that you used to test your explanation.

The regression method is a useful tool for describing the relationship between two variables when we think that one variable depends on the other in a linear manner. Regression can then be used to predict values for the response when we know a predictor variable. In this lab we will explore the relationship between variables in the 1999 cars data set.

The regression output shown at the right are examples of the output supplied by Data Desk and CrunchIt. These outputs show how a car's fuel efficiency in city driving, 'mpg:city', depends on the horsepower of the car's engine.

```
Dependent variable is:    mpg:city
No Selector
R squared = 48.4%     R squared (adjusted) = 48.1%
s =  3.867  with  140 - 2 = 138  degrees of freedom

Source        Sum of Squares    df    Mean Square    F-ratio
Regression    1938.74            1     1938.74        130
Residual      2063.48          138     14.9527

Variable      Coefficient    s.e. of Coeff    t-ratio     prob
Constant      32.2008        1                32.2       ≤ 0.0001
horsepower    -0.057183      0.005022        -11.4       ≤ 0.0001
```

Examine one of the regression output windows and find the values of the coefficients and the dependent variable (which we call the y-variable). In the window above, 'mpg:city' is the dependent variable, the constant term in the model is 32.20082 (that's the estimate of y when x = 0), and the coefficient for horsepower (the x-variable) is −0.05183. The coefficient for horsepower tells us that for each increase of one horsepower, we expect that the miles per gallon will decrease by 0.057183. The negative coefficient indicates a downward slope.

We can see the negative linear relationship between these variables also by looking at the scatterplot.

Software Tip: Creating Regression Output

In Data Desk select the variable of interest (place the Y on this variable). Hold the shift key while selecting the independent variable (an X will be placed on the variable. Choose **Calc>Regression.**

CrunchIt! In CrunchIt click **Stat>Regression>Simple Linear** Choose the X and Y variables of interest.

AT THE COMPUTER

We can further explore the relationship between variables in the 1999 Cars data file.

1. Open the data file 'Cars99' and produce the regression output for predicting the city gas mileage (mpg:city) using the weight (thousands of pounds) of the car. What is the regression equation that predicts the city gas mileage using weight?

$$\text{Mpg:city} = \underline{\hspace{1cm}} + \underline{\hspace{1cm}} *\text{weight}$$

2. Explain in layman's terms what the slope term means in this equation.

3. We would like to predict the city gas mileage of a car that weighs 3000 pounds. Use the equation you produced in the previous problem to make this prediction.

4. What is the equation that predicts the city gas mileage using the engine displacement of a car?

$$\text{Mpg:city} = \underline{\hspace{1cm}} + \underline{\hspace{1cm}} *\text{displacement}$$

5. Explain in layman's terms what the slope term means in this equation.

The regression output also shows the coefficient of determination (R^2), which is the square of the correlation coefficient. The value of R^2 can be interpreted as the percent of the variability in 'mpg:city' explained by knowing the independent variable. For example, the horsepower of a car explains 48.4% of the variability in city gas mileage. (The remaining 51.6% of the variability is from other factors such as aerodynamics of the car, the weight of the car, etc.) The value of R^2 ranges from 0 to 100% and is often used to compare different predictor variables.

6. Examine the regression outputs for predicting city gas mileage using weight and displacement. Which of these predictor variables produces the highest R^2?

IN THE LAB

Next we will conduct a study in which we will use regression to predict the amount of *time* it takes to complete a "connect the dots" maze based on the number of dots in the maze.

Outline of study

    Some members of the class will serve as the timers and will use a stopwatch to time how long it takes each student to complete the maze.

    Volunteers from your class will each complete a "connect-the-dot" maze (available from your instructor). The different mazes will vary in length.

    The resulting times needed to complete the mazes for all of the participating students will be recorded, along with the corresponding number of dots in the maze.

Other questions the class should consider prior to starting the experiment

    How is the timing to be done? For example, how is it to be determined when the subject has completed the maze? How should the maze be started?

    How should the mazes be selected to assign to each subject?

    Under what conditions should the mazes be completed? For example, should others be watching?

    What other precautions should be taken to ensure that the experiment gives a fair representation of the relationship being studied?

When the data has been collected use computer software to answer these questions.

7. Make a scatterplot of 'time' versus 'dots'. Does the relationship appear to be linear? Explain.

8. Fit a Regression line to examine the relationship between the time it takes to complete the maze and the length of the maze. What is the fitted regression line?

Estimate of time to complete the maze = _____ + _____ * (number of dots)

9. If a student completed a maze with 30 dots, on average how much time do you predict that student took to complete the maze?

10. About what percent of the variation in the time is explained by knowing the number of dots? Explain.

11. Would the regression method be as accurate for predicting the time needed to complete a maze with 100 dots? Explain.

LAB 10
Exploring the Law of Averages

A mutual fund allows investors to simultaneously own many stocks and collect earnings according to the average gains of the stocks in the mutual funds' portfolio. People pay a small premium for this choice of stock ownership in order to avoid the wild fluctuations that are possible with an individual stock. The volatility of individual events and the stability of averages is the key to the Law of Averages that we will study in this lab. Tracking the fluctuations in the value of individual stocks, mutual funds, or stock indices is easy to do using the charting facilities at the Web sites of the major stock exchanges (e.g., www.nasdaq.com, www.amex.com, or www.nyse.com). There you can see the Law of Averages at work by charting an index like the Dow Jones Industrial Average and comparing the degree to which it fluctuates with the fluctuations in the price of the individual Dow stocks.

The Law of Averages describes the behavior of experimental results as the number of independent trials increases (e.g., as we add more flips of a coin or more spins of a roulette wheel).

> **The Law of Averages:** Averages or proportions are likely to be more stable when there are more trials, whereas sums or counts are likely to be more variable. This does not happen by compensation for a bad run of luck because independent trials have no memory.

When an American roulette wheel is spun, the ball lands randomly in one of 38 different pockets. Two pockets are colored green and labeled "0" and "00." The remaining are labeled with the numbers from 1 through 36 with eighteen numbers colored red (1, 3, 5, 7, 9, 12, 14, 16, 18, 19, 21, 23, 25, 27, 30, 32, 34, 36) and eighteen numbers colored black (2, 4, 6, 8, 10, 11, 13, 15, 17, 20, 22, 24, 26, 28, 29, 31, 33, 35).

1. If you bet that a red number will come up, what is your chance of winning?

2. If you bet a dollar on red, the casino will give you a dollar if the ball lands on red and take your dollar if it lands on black or green. Consider what the Law of Averages says about the proportion of times that the ball will land on red if the wheel is spun over and over again. Write this in your own words below.

3. When are you more likely to come out ahead (i.e., win more than 50% of your bets) – if you bet 38 times on red or if you bet 3800 times on red? Explain.

IN THE LAB

Using the Law of Averages requires that we can recognize the basic components of an experimental process. What are the individual "trials" of the process? How many trials are involved in the experiment? Are the trials independent? What is the chance behavior for one trial? How does the final response variable relate to the individual trials?

Once these raw components are recognized, the chance behavior of the experimental results are determined. For example, spinning a roulette wheel 38 times and counting how many times it comes up red involves 38 trials representing the 38 spins. The roulette wheel has no memory so the trials are independent. On each spin you have 18 chances out of 38 to get a red. The final response variable tracks whether each trial resulted in a red (or not) and keeps a running count that goes up by one for each red. Suppose instead that we have a bag with 38 marbles, including 18 red marbles and 20 white marbles. Consider the experiment of drawing 38 times from the bag and counting how many red marbles we pick (each time replacing the marble we get). The raw components are all the same as 38 roulette bets on red: 38 independent trials with 18 chances out of 38 on each trial of getting a red. Thus the behavior of the process must be the same.

4. Draw 38 times from the bag of 38 marbles provided by your instructor. How many times did you get a "red" result in your 38 trials? How many did you expect?

5. For each student in class today, record the number of times that red came up in their 38 draws from a bag of 18 red and 20 white marbles.

___ ___ ___ ___ ___ ___ ___ ___ ___ ___ ___ ___

___ ___ ___ ___ ___ ___ ___ ___ ___ ___ ___ ___

___ ___ ___ ___ ___ ___ ___ ___ ___ ___ ___ ___

6. Most of these values are within plus or minus _____ of the expected number. (Fill in the blank with a number.)

7. How many of these experiments ended up with exactly the number of reds expected?

8. Record the results above as a percentage of the 38 trials (e.g., if you saw red come up 21 times for one experiment record $100(21/38)\% = 55.26\%$ in that case).

___ ___ ___ ___ ___ ___ ___ ___ ___ ___ ___ ___

___ ___ ___ ___ ___ ___ ___ ___ ___ ___ ___ ___

___ ___ ___ ___ ___ ___ ___ ___ ___ ___ ___ ___

9. How much did these values fluctuate around the percentage of time you expected a red to come up?

AT THE COMPUTER
We could repeat the exercise for 3800 trials; however, that would take a substantial amount of time pulling marbles from the bag. You can also have the computer perform these kinds of experiments. Independent replications of a situation that has two possible outcomes (e.g., succeed/fail, yes/no, win/lose, heads/tails, etc...) are called "Bernoulli trials." You can have your software perform a sequence of Bernoulli trials, such as winning or losing when you bet on red at roulette. You just have to provide information about the components of the process. For example, to simulate the results of 38 spins of the roulette wheel you should specify 38 spins and a probability of success equal to 18/38 ≈ 0.47368. The software will create a new variable which will have a value of one to

represent the outcome you are tracking (red) and a value of zero for anything else. You can then summarize the results to get an idea of how many trials were red.

Software Tip: Simulating Bernoulli Trials

In Data Desk, click on **Manip>Generate Random Numbers.** A window will appear in which you can select "Bernoulli trials." Specify a probability of success and the number of random trials you want.

Generate Random Data

Generate 1 variables with 38 cases.

Distribution:
- ○ Uniform
- ○ Normal mu = _____ sigma = _____
- ⦿ Bernoulli trials: Prob(success) 0.47368
- ○ Binomial experiments:
    - # Bernoulli trials / experiment _____
    - Probability (success) = _____
- ○ Poisson: lambda = _____

Seed = _____

Generator [Linear Congruential ▼]

[OK] [Cancel] [Help]

CrunchIt!

In CrunchIt click **Data>Simulate Data>Bernoulli.** Enter the number of columns, rows, and the probability of a success on an individual trial.

Bernoulli samples

Rows:	38
Columns:	1
p:	0.47368

[?] [Cancel] [Simulate]

Java Applet Window

Software Tip: Frequency Tables

Click on the variable of interest, then click **Calc>Frequency Breakdowns**.

CrunchIt!

In CrunchIt click **Stat>Tables>Frequency** and enter the name of the variable.

---

10. How many times did the software simulate a "red" result in your 38 trials? How many did you expect? Did any one in your class get more than 10 away from what you expect?

11. Now have your software perform 3800 Bernoulli trials that simulate spinning a roulette wheel 3800 times and counting how many reds come up. Record your results and the results of your classmates below.

___ ___ ___ ___ ___ ___ ___ ___ ___ ___ ___ ___ ___

___ ___ ___ ___ ___ ___ ___ ___ ___ ___ ___ ___ ___

___ ___ ___ ___ ___ ___ ___ ___ ___ ___ ___ ___ ___

12. How much did these values fluctuate around the number of reds you expect to happen in 3800 spins? How many of these experiments ended up with exactly the number of reds expected? How many were more than 10 away from what you expect?

13. How do these results compare with your results for the counts when we had only 38 simulated "spins" based on drawing marbles out of a bag or using your software?

14. Explain how the counts you found in the 38-spin trials and the 3800-spin trials illustrate the Law of Averages?

15. Record the results from above as a percentage of the 3800 spins.

___ ___ ___ ___ ___ ___ ___ ___ ___ ___ ___ ___ ___

___ ___ ___ ___ ___ ___ ___ ___ ___ ___ ___ ___ ___

___ ___ ___ ___ ___ ___ ___ ___ ___ ___ ___ ___ ___

16. How much did these values fluctuate around the percentage of time you expect a red to come up?

17. How do these results compare with your results for the percentages in Question 9 above? Does this comparison illustrate the Law of Averages? Explain.

18. In how many of the experiments with 38 trials did you get a majority of reds (i.e., more than 19)? In how many of the experiments with 3800 trials did you get a majority of reds (i.e., more than 1900)? Does this agree with what you predicted back in Question 3? Explain.

LAB 11
Normal Approximation

The Normal distribution has a histogram that looks like a bell-shaped curve. Here's why it is important in the theory of sampling distributions:

> **The Normal Approximation:** Averages or proportions based on a large number of independent trials of an experiment closely follow the normal distribution.

In this lab we investigate the quality of the normal approximation to the sampling distribution of an average or proportion. The quality of the approximation depends on how close the histogram of the population values is to following a normal curve and on the size of the sample. We will use the computer to investigate the performance of the normal approximation for proportions.

DICHOTOMOUS POPULATIONS

Situations in which we can group the population into two groups (like when we are asking a yes/no question) are referred to as dichotomous or binomial populations. For sampling from a dichotomous population, a good rule to follow is that it is not appropriate to use the normal approximation for proportions unless the expected count (np) *and* the expected count of the opposite outcome ($n(1-p)$) are both at least five. Other statisticians suggest that the normal approximation is good for most purposes as long as the standard deviation of the sampling distribution of the count ($\sqrt{np(1-p)}$) is greater than two.

Both rules illustrate the theme that the normal approximation does better when the sample size, n, is larger and when the chance of the outcome you are tracking, p, is closer to one-half.

We will examine the normal approximation for the proportion of times something happens in four situations:

Drawing 10 times ($n = 10$) from a dichotomous population

- with a 50% chance of the outcome you are tracking ($p = 0.5$)
- with a 90% chance of the outcome you are tracking ($p = 0.9$)

Drawing 40 times ($n = 40$) from a dichotomous population

- with a 50% chance of the outcome you are tracking ($p = 0.5$)
- with a 90% chance of the outcome you are tracking ($p = 0.9$)

1. Based on your studies of the normal approximation, write down your guess for which of these four sums will come closest to following the normal curve. Explain your choices briefly.

    a. Which normal approximation will work best?

    b. Which normal approximation will work most poorly?

    c. Which normal approximations will fall in between?

AT THE COMPUTER

In lab 10 on the Law of Averages lab we learned how Bernoulli trials could be simulated. A Binomial experiment is a series of Bernoulli trials, where we count the total number of successes. Use your computer software to make 1000 repetitions of each of the four binomial experiments described above. Be sure you keep track of which situation is represented by each variable.

Software Tip: Generating Binomial Experiments

CrunchIt!

In CrunchIt click **Data>Simulate Data>Binomial.** In the dialogue box the number of rows will be the number of random numbers you want to generate, the columns will generally be 1. Fill in the number of draws and the probability.

Binomial samples

Rows: 1000

Columns: 1

n: 10

p: 0.5

?   Cancel   Simulate

Java Applet Window

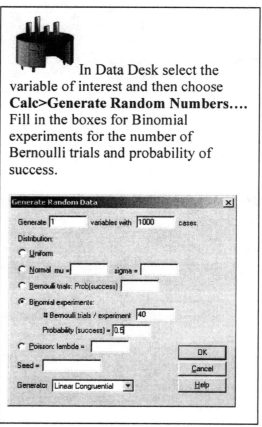

In Data Desk select the variable of interest and then choose **Calc>Generate Random Numbers....** Fill in the boxes for Binomial experiments for the number of Bernoulli trials and probability of success.

Generate Random Data

Generate 1   variables with 1000   cases.

Distribution:

○ Uniform

○ Normal mu =   sigma =

○ Bernoulli trials: Prob(success)

● Binomial experiments:

    # Bernoulli trials / experiment 40

    Probability (success) = 0.5

○ Poisson: lambda =

Seed =

Generator: Linear Congruential

OK   Cancel   Help

2. Make histograms of each of the four variables with bin widths of 1. First examine the two 10-draw experiments. Sketch these two observed histograms below and describe their shapes. These histograms reflect the number of times (out of 10) that the outcome you are tracking occurred. If you divided these variables by 10, so they would now represent the proportion of times that the outcome occurred, how would this affect their histograms? (Explain how it would affect their center, spread, and shape.)

3. Did the populations with 50% chances or the populations with 90% chances produce a histogram that looks more like the normal curve? Does this correspond with what you expected?

4. Next, examine the two 40-draw experiments.
    a. Sketch both of these histograms below.

    b. Did the population with 50% chances or the populations with 90% chances produce a histogram that looks more like the normal curve in this instance? Comment.

    c. How do these two histograms compare with the histograms for the 10-draw experiments with the same chances? Comment.

LAB 12
Confidence Intervals

The Mars candy company makes plain m&m's® candies in red, yellow, orange, brown, green, and blue and randomly mixes the colors according to a specific percentage distribution. In this activity we'll investigate the proportion of m&m's that are blue. To get an estimate of this proportion, we need to take a sample. But the sample proportion is subject to chance variability and we can express this by providing a confidence interval. The idea that each student will be dealing with an independent random sample provides the key to the interpretation of these confidence statements, as you'll see in this lab.

1. Your instructor will provide the class with a large quantity of plain m&m's. Decide as a class how you can obtain a sample that would not be biased. Are there any problems with the procedure that were difficult to resolve?

2. Define the population of interest and the parameter p being estimated. Explain briefly.

3. Carefully use the procedure you decided on to count out 40 candies. Count the number of blue m&m's that you obtained. I found _____ blue m&m's in my sample. As a fraction of the sample size this gives $\hat{p} =$ _____.

4. Estimate the standard deviation for the proportion of blue m&m's using your sample.

The variability in your sample proportion of blue m&m's can be described as

Sample proportion = population proportion + random variation

where the random variation from sample to sample approximately follows a normal distribution (Why?).

When all of the students in class today have estimated the percentage of blue m&m's, we can look at the question we started with: What is the true percentage of blue candies among all m&m's?

The size of the random variation will be less than one standard deviation of $\hat{p}$ about 68% of the time, so the interval

sample proportion ± 1*estimated standard deviation of $\hat{p}$

should contain the population proportion for about 68% of the samples. This is an example of a *confidence interval*.

5. My 68% confidence interval is _____ ± _____ , which goes from a proportion of _____ up to _____ blue.

There is a trade-off between confidence and precision. You can have more confidence in an interval estimate if you are willing to be less precise. For example, because the size of the random variation is less than two standard deviations of $\hat{p}$ for about 95% of the samples, we can make a 95% confidence interval using:

sample proportion ± 1.96*(estimated standard deviation of $\hat{p}$)

6. A 95% confidence interval for the proportion of blue m&m's based on my sample would go from _____ up to _____ .

7. Write your answers in the appropriate location on the board as your instructor indicates. Different samples were drawn by each student, and they each constructed their own confidence interval. List this information in the table on the following page.

According to the Department of Consumer Information of Mars, Inc., m&m's plain candies are randomly mixed to contain 24% blue candies. Did the 68% confidence interval that you constructed contain the $p = 0.24$ figure given by the company?

8. For all of the students in class today, how many of their 68% confidence intervals included the $p = 0.24$ figure supplied by the company? How does this compare with what you would expect? Explain.

9. How many of the 95% confidence intervals that were made by the students in class today include the company figure of 0.24? Is this close to what you expected? Explain.

Sample number	Number blue	$\hat{p}$	Beginning of 68% interval	End of 68% interval	Beginning of 95% interval	End of 95% interval
1						
2						
3						
4						
5						
6						
7						
8						
9						
10						
11						
12						
13						
14						
15						
16						
17						
18						
19						
20						
21						
22						
23						
24						
25						
26						
27						
28						
29						
30						

Remember, to interpret a confidence interval, you have to think about what would happen if the sampling procedure were repeated over and over again.

EVERY HOMEOWNER'S NIGHTMARE

Rick Miller '92

$\mathbf{A}$n optometrist studies the comfort of a new type of contact lens by having subjects wear a standard lens in one eye and the new type of contact in the other. She later counts how many subjects found the new contact to be more comfortable. A virologist studies the effectiveness of a new flu vaccine using twins living apart – randomly assigning one twin to get the vaccine and one to a placebo. He later counts how many times the twin on treatment showed fewer flu symptoms than the twin on placebo. These are examples of situations where the units of observation in an experiment are really pairs. In this lab we will analyze paired data from an experiment on the stress of classroom participation carried out on 99 statistics students at Ohio State University in 2001. We will also use computer software to simulate the results of a hypothesis test.

THE CLASSROOM STRESS PROTOCOL:

The purpose of that experiment was to see if a student's blood pressure rises when they are called on to speak in class. Here is an outline of how it was conducted:

Student volunteers gathered in a mock classroom situation and a few minutes into a "lecture" they took their own blood pressure to serve as a baseline value.

During the remainder of the "lecture," the instructor randomly called on students to answer questions.

Immediately after responding, students took their blood pressure again.

At the end of the experiment we count how many students' blood pressure was higher after speaking versus how many had lower blood pressure after speaking.

A HYPOTHESIS TEST

1. If we assume that speaking does not affect blood pressure, then the chance that an individual student's pressure goes up just by chance would be _____. In other words, we would we have a null

    $H_0: p =$ _____

2. On the other hand, if speaking in class made it more likely that your blood pressure would rise, we would write this alternative hypothesis as

    $H_a:$

3. In our data we have a sample of $n = 99$ students. If we assume that the null hypothesis is true, we would expect a sampling distribution that is approximately normally distributed and is centered at a mean of _____ and a standard deviation that is given by

$$\sqrt{\frac{p(1-p)}{n}} = \underline{\hspace{2cm}}$$

4. In the actual experiment, it turned out that blood pressure went up for 66 students and went down for 38 students after speaking in the mock classroom situation. Thus, the sample gives an estimated probability of $\hat{p} = \underline{\hspace{1cm}}$.

5. From this sample value we can calculate the standard score by using the sampling distribution and calculating:

$$\frac{\text{observation - mean}}{\text{standard deviation}} =$$

6. Using table B or the normal curve density applet, we see that the probability value (p-value) corresponding to this standard score would be _____.

7. What do you conclude about the null hypothesis based on this p-value?

WHAT DOES A P-VALUE MEAN?
Could the results you found be just by random chance? What does the probability you found in question 6 mean? Let's examine a simulation that tells us about this scenario.

8. We begin by assuming that the proportion of students whose blood pressure would go up is 0.5 (50%). If this proportion was 0.5, then the sample proportion we see should be near 0.5. The variability should be similar to what we would see in 99 flips of a fair coin. We can simulate these results by using Bernoulli trials (See the Lab 10). In this case simulate 99 Bernoulli *trials* with a *probability of 0.5*, and use a frequency table to summarize the proportion of 1's. What proportion were 1's? In other words, the value of $\hat{p} = \underline{\hspace{1cm}}$.

9. Calculate the test statistic for your simulated sample proportion.

$$\frac{\text{observation - mean}}{\text{standard deviation}} = \underline{\hspace{2cm}}$$

10. Repeat this simulation nine more times for a total of 10 simulations, and record the *test statistic* for each.

_____   _____   _____   _____   _____   _____   _____   _____   _____   _____

11. If blood pressure is not affected by speaking in class, then we would expect your test statistics to be around zero. How many of your simulated values were farther from zero than the actual test statistic you saw in question 5?

12. Your instructor will ask you to record your values from the test statistic on the board. In how many of the class' simulated samples did you see a test statistic that was farther from zero than what you saw in question 5?

13. In your class' simulation _____ out of _____ or _____% of simulated samples produced a test statistic as extreme or more extreme than the actual sample we saw.

14. Examine the probability value you found in question 6. Use the results of your class' simulations to explain in layman's terms what the probability value means.

LAB 14
Testing a Numeric Hypothesis

A study by Dr. John Manning claims that the ring finger on a person's hand is *typically longer* than their index finger. In this activity we will test his idea and determine if there is evidence that this is true.

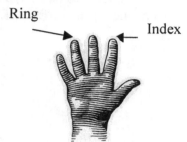

Ring

Index

The Study: In this activity we will record the length of the index and ring fingers for the students in your class. Next we will take the length of the ring finger minus the length of the index finger.

1. Assume ring fingers are NOT really longer than index fingers and that any variability we see is just by random chance. On average you would expect the difference (ring – index) in finger length to be around _____.

2. If Dr. Manning is correct, we would expect the difference (ring –index) to be _____.

3. Discuss as a class how you will measure the fingers to get a reliable and unbiased measure. Record the finger lengths of the student volunteers from your class. Calculate the difference in finger length.

Student	Ring	Index	(Ring – Index)	Student	Ring	Index	(Ring – Index)
1				15			
2				16			
3				17			
4				18			
5				19			
6				20			
7				21			
8				22			
9				23			
10				24			
11				25			
12				26			
13				27			
14				28			

4. What is the average difference in ring and index finger length? _____

5. What is the standard deviation of these differences? (Use a calculator or computer to find this value.) _____

6. If there is really *no* difference between ring and index finger length could your answer in question 4 have occurred just by chance? Conduct a test of hypothesis. (Remember that your null hypothesis should assume there is no difference.)

    a. Your hypotheses:
       H_0:

       H_a:

    b. Your standard score: _____

    c. The p-value:_____

    d. What is your conclusion?

7. Explain the meaning of your conclusion in practical terms. Is the average you found in your sample significantly different from what we would expect if there is no difference in finger length?

8. Dr. Manning's original work dealt only with the finger lengths of men. What null and alternative hypotheses would be examined to determine if men and women were different in terms of this ring and index finger comparison?

LAB 15
Testing Relationships with the Chi-square

The chi-square test of independence is one of the most widely used tests of hypothesis. It allows us to examine the relationship between two categorical variables. Is there a relationship between the gender of a student and if the student is registered to vote? Or whether the student has taken an airline flight? In this activity we will examine the relationship between several categorical variables present in a survey of students and see how the chi-square test of hypothesis works.

AT THE COMPUTER

1. Open the data file 'WI02Survey'. This file has several categorical variables, including the gender of the student. How does gender relate to whether or not students ate breakfast the day of the survey? Create a two-way table that examines the relationship between gender and if the student has eaten breakfast. Fill in the values below.

	Didn't eat	Ate breakfast	Total
Females			314
Males			188
Total	233	269	502

2. What proportion of the students in that class was female? _____

To formally test this hypothesis we will begin by establishing hypotheses. To conduct this test we will assume that gender and eating breakfast are not related. More formally:

H_0: There is no association between gender and eating breakfast.

Another way to think of this would be that the proportion of females would be the same in both columns of the table. On the other hand, the alternative hypothesis would be that the variables are related, *or*

H_a: There is an association between gender and eating breakfast.

55

3. As with other tests of hypothesis we will assume the null hypothesis is true in calculating our test statistic and p-value. If the null hypothesis is true how many subjects would we expect in each cell of the table. To determine this we need only think about the fact that we would expect the same proportion of females (and males) in each column of the table.

   a. There are 233 students who did not eat breakfast. Multiply the proportion of students who were female that you found in question 2 by 233 (keep these values to two decimal places).

   Expected count = Proportion female * total who didn't eat breakfast = _____

   This calculation is equivalent to finding expected count $= \dfrac{\text{row total * column total}}{\text{table total}}$

   b. Repeat this calculation for the other cells of the table:
   Expected count = Proportion female * total who ate breakfast = _____
   Expected count = Proportion male * total who didn't eat breakfast = _____
   Expected count = Proportion male * total who ate breakfast = _____

   c. Enter these values into the table below. How do the expected values compare with the actual values you found from the data?

	Didn't eat	Ate breakfast	Total
Females			314
Males			188
Total	233	269	502

4. We can measure the actual difference between the observed and the expected values by calculating the statistic of the form:

$$X^2 = \sum \frac{(\text{observed count - expected count})^2}{\text{expected count}}$$

$$= \frac{(\underline{} - \underline{})^2}{\underline{}} + \frac{(\underline{} - \underline{})^2}{\underline{}} + \frac{(\underline{} - \underline{})^2}{\underline{}} + \frac{(\underline{} - \underline{})^2}{\underline{}}.$$

5. Is this test statistic one that would be unusual or could this arrangement have happened just by chance? We can determine this by comparing the test statistic with the sampling distribution we would expect. This sampling distribution is the chi-square distribution with 1 degree of freedom. Use the table of the chi-square to determine approximately the p-value of this test.

   p-value = _____

6. What conclusion would you make in this setting? Explain what your conclusion would mean in terms of the relationship between eating breakfast and gender.

The calculations you have done in the previous questions can be automated by the computer software. Both CrunchIt and Data Desk can calculate the chi-square statistic and the p-value. Examine the output and see how these are presented.

Software Tip: Calculating the Chi-square	
CrunchIt! In CrunchIt the chi-square statistic is calculated with the two-way table automatically.	In Data Desk create a two-way table for the variables of interest. Click on the hyper view triangle and choose 'Table Options'. In the dialogue box click **chi-square value.**

7. Use the software to create the two-way table and calculate the chi-square statistic for the relationship between the gender of the student and whether the student has taken a flight on a commercial airline in the last 30 days.
   a. Test statistic X^2 = _____

   b. P-value = _____

   c. Use your p-value to make a conclusion about the relationship of these variables.

8. Use software to examine the relationship between gender and whether the student had seen a play in the last six months.
   a. Test statistic $X^2 =$ _____

   b. P-value = _____

   c. Use your p-value to make a conclusion about the relationship of these variables.

9. Use software to examine the relationship between gender and whether the student is registered to vote.
   a. Test statistic $X^2 =$ _____

   b. P-value = _____

   c. Use your p-value to make a conclusion about the relationship of these variables.

10. Which of the four relationships you examined were significant? Can we conclude there is a cause-and-effect relationship between gender and the variables you examined? Explain why or why not.

11. Can the sample of students who answered the survey analyzed in this lab be viewed as a random sample from some population? How does your answer to this question affect the conclusions you can draw from your significance tests?

Additional Problems for Chapters 1 - 4

1. The Division of Traffic and Parking at a large university wants to know the percentage of students who used the campus bus system last week. They take a random sample of four hundred students from their list of students with parking stickers and find that 20% of them have used the bus system during the past week. The results of this survey are reported as follows: "With 95% confidence, the percentage of students using the bus system lies in the range 20% ± 4%."

   a. Identify each of the following:

   parameter   _____

   population   _____

   sample   _____

   sample statistic _____

   level of confidence _____

   margin of error   _____

   sampling frame   _____

   b. This survey is likely to be biased because:
      i) the sampling frame is not the same as the population.
      ii) the margin of error is too large.
      iii) of nonresponse bias.

   Pick one and briefly explain (including a statement of whether the problem is a type of sampling error or a type of non-sampling error).

2. A random sample of 900 people was taken in Chicago to estimate the percentage of Chicago voters that favor an end to the U.S. embargo against Cuba. The results of this survey are given as a 95% confidence interval that turns out to be 53% ± 3.4%.

   a. If we had calculated a 99% confidence interval for this poll, how would the margin of error have been different? Explain.

   b. A second poll, using the same methods, is taken of 900 people in Columbus, Ohio, which has about one-fourth as many voters as Chicago. Will the margin of error for the Columbus poll be larger than 3.4%, smaller than 3.4%, or about 3.4%? Explain.

3. The Denver Public Library wishes to estimate the percentage of Denver households with an adult who has read at least one book in the last month. The homes of four hundred customers who have library cards are sampled and it turns out that 90% of these households have an adult who has read a book in the past month. Is 90% likely to be a biased estimate for the true percentage of Denver households with an adult who has read at least one book in the last month? Explain why or why not.

4. On Saturday, October 15, 1997, broadcast and print media across the country reported on the government's release of a report on the status of working women called "Working Women Count." This report gives findings from a survey of 250,000 women. According to one news report:

   "More than 1600 businesses, unions, newspapers, magazines, and community service organizations helped distribute the survey to their members, subscribers, and patrons, which the White House announced with much fanfare. It sought women's opinions on job satisfaction, pay, benefits, and opportunities for advancement."

   A second survey asking the same questions was conducted at the same time. However this survey interviewed only 1200 working women chosen at random.

   Which survey is likely to give a more accurate view of working "women's opinions on job satisfaction, pay, benefits, and opportunities for advancement"? Is this improved accuracy likely to be the result of lower bias or smaller sampling variability? Explain briefly.

5. Suppose you want to know the average amount of money spent on concessions by the fans attending opening day for the Cleveland Indians baseball season. You get permission from the team's management to conduct a survey at the stadium, but they will not allow you to bother the fans in the club seating or box seat areas (the most expensive seating, where the fans have paid over $30 per ticket). Using a computer, you randomly select 500 seats from the rest of the stadium and during the game ask the fans in those seats how much they spent that day. Provide a reason why this survey might yield a biased result. Explain whether the reason you provide is a type of sampling error or a type of non-sampling error.

6. On Wednesday, October 5, 1997, the WBNS evening news covered the story of a plan to build a light rail public transportation system financed by a half-cent increase in the sales tax. After the story was broadcast, the news anchor said that the station was interested in the opinions of the viewing audience, and invited them to register their opinions by calling one of two "900" phone numbers. Calling one number would register their opinion as favoring the proposal, whereas calling the other number would register their opinion as being against the proposal. Each call cost $0.50. The results of this survey were reported on the Thursday night broadcast.

   a. Give one reason why the results are likely to be biased. In which direction do you think this bias is likely to go?

   b. State whether your reason is an example of sampling error or non-sampling error.

7. Suppose the mayor of Los Angeles wants to sample the opinions of the adult residents of the city in order to estimate the percentage who favor the mayor's proposal to tighten restrictions on public smoking. Nine hundred people are chosen at random from the Los Angeles phone directory, and 60% of them favor the mayor's proposal. The survey research firm hired by the mayor's office then reports the results as follows: "With 95% confidence, the percentage favoring the new restrictions lies in the range 60% ± 3%."

   a. Identify each of the following,

parameter	_____
population	_____
sample	_____
sample statistic	_____
level of confidence	_____
margin of error	_____
sampling frame	_____

   b. This survey may be biased because:
      i) the sampling frame is not the same as the population.
      ii) the margin of error is too large.
      iii) of nonresponse bias.

   Pick one and briefly explain (including a statement of whether the problem is a type of sampling error or a type of non-sampling error).

8. A survey is conducted in Chicago (population 2,800,000) using random-digit-dialing equipment that places calls at random to residential phones, both listed and unlisted. The purpose of the survey is to determine the percentage of adults who would favor a half-cent increase in the sales tax to help fund public transportation. Four hundred adults are interviewed and 36% of them favor the proposal. A second survey is taken in Dayton, Ohio, (population 180,000) using the same techniques and asking the same question of 400 adults living in Dayton.

   a. For the Chicago survey, identify the population, the sampling frame, the sample, the variable measured, the parameter of interest, and the corresponding statistic.

   b. The margin of error of the Chicago survey will be:
      i) greater than the margin of error of the Dayton survey.
      ii) about the same as the margin of error of the Dayton survey.
      iii) less than the margin of error of the Dayton survey.

   Pick one and explain briefly.

9. As part of a class project, each student in a statistics class at a large university drew names at random from the student directory and asked the people in their sample if they had a campus parking sticker. The statistics students were required to draw a sample of 50 names, but extra credit was given for randomly selecting and interviewing an additional 50 names (i.e., a total random sample of 100). Everyone who was surveyed responded to the question. Finally, a check with the university's Division of Traffic and Parking showed that 40% of all students were issued a parking sticker. Which of the following are true and which are false?

    a. If the directory doesn't list every student at the university, then these student polls have a possible source of sampling error.

    b. If the directory doesn't list every student at the university, then all of these student polls will have comparable levels of precision in their estimates – even those who got the extra credit.

    c. If the directory doesn't list every student at the university, then all of these student polls will have comparable levels of bias in their estimates – even those who got the extra credit.

    d. Assuming that the directory lists every student at the university, all of the individual surveys will get exactly 40% as their estimate of the percentage of students with parking stickers.

    e. Assuming that the directory lists every student at the university and each statistics student creates a 95% confidence statement, then we'd expect all of these confidence statements to include the 40% value given by the Division of Traffic and Parking.

    f. If each statistics student creates a 95% confidence statement, then the margin of error will be smaller in the confidence statements made by the group who got the extra credit.

10. A random sample of 1000 people who signed a card saying they intended to quit smoking on November 20, 1995 (the day of the "Great American Smoke-out") were contacted in June 1996. It turned out that 210 (21%) of the sampled individuals had not smoked over the previous 6 months.

    a. Specify the population of interest, the parameter of interest, the sample, and the sample statistic in this problem.

    b. Use the quick method to estimate the margin of error in this situation.

11. A researcher is interested in the number of hours parents of pre-school children in the United States spend reading to their children. To investigate this, he obtains the student list from a local preschool and contacts the parents of 30 children from this list. The parents reported an average of 8.2 hours per week with a standard deviation of 3 hours.

   a. What is the population of interest in this problem?

   b. What is the parameter of interest?

   c. Another researcher offered the following criticism of this study: "The sample is biased because the students all come from the same school and the sample does not represent the entire population of interest." Would the problem described by this researcher be a sampling or a non-sampling error? Explain.

12. Consider the following report of a nationwide survey on Gallup.com.

PRINCETON, NJ – A recent Gallup Youth Survey presented American teenagers (aged 13 to 17) with two questions about the death penalty. One asked if they personally believed it was morally acceptable; the other asked them to choose whether death or life imprisonment should be the penalty for murder. The results indicate that more than three in five (62%) believe that the death penalty is a morally acceptable option, and slightly more than half (52%) believe the death penalty is the better punishment for murder. Teenage boys express more conservative stances on capital punishment than teenage girls do. Boys are more likely to favor the death penalty as punishment for murder (59% to 45%), and to consider the death penalty to be morally acceptable (67% vs 57%). The questionnaire was completed by 517 respondents. For results based on the total sample, one can say with 95% confidence that the margin of sampling error is ± 5%.

   If the Gallup Poll was interested in just New Jersey youth rather than all American teenagers, but still questioned 517 respondents, would the reliability of the poll improve, worsen, or not change much? Pick one and explain.

EESEE Exercises for Chapters 1 - 4

The following exercises make use of stories in the Electronic Encyclopedia of Statistics Examples and Exercises, or EESEE (pronounced ee-zee). EESEE is included in the E-STAT Pack that accompanies this workbook. You can also access EESEE at www.whfreeman.com/eesee.

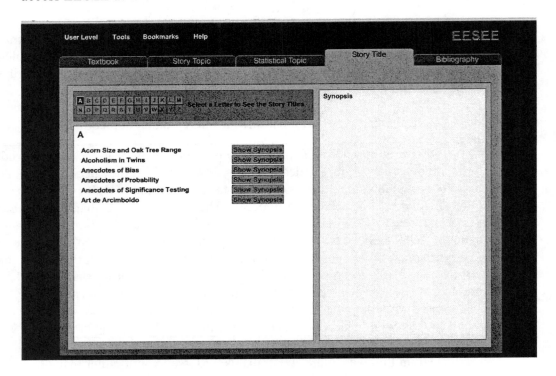

You may find specific stories by looking for the title on the "Story Title" tab.

13. **EESEE story** *Nutrition and Breakfast Cereals.* American children grow up with breakfast cereals. Many children's television programs are sponsored by breakfast cereal companies. Not surprisingly, cereals are also a staple of college student breakfasts. What are the nutritional characteristics of breakfast cereals? The EESEE story "Nutrition and Breakfast Cereals" examines this question. Open the story and read through the introduction and also click on the **Data** tab to review the list of variables.

   a. What are the individuals in this statistical study? How many individuals are included?

   b. How many variables are included in this study? Which variables take numeric values?

14. **EESEE story** *Perot Poll* (Note: This is the same poll discussed in Example 6 in Chapter 4 of the text.)

   a. What is the population of interest for Perot's *TV Guide* survey? Provide two reasons why the people who responded to the Perot survey are not likely to be representative of this population. Explain each briefly.

   b. Which survey is likely to provide more accurate information about the opinions of the population of interest regarding campaign finance reform, the Perot survey or the Yankelovich survey? Is the survey you chose likely to be more accurate because of increased precision (i.e., better reliability) or because of decreased bias? Explain briefly.

15. **EESEE story** *Anecdotes of Bias.* In 1992, flight attendants handed out a survey to passengers who were flying from New York to Chicago on Midway Airlines. Based on the results of this survey, Midway Airlines purchased advertisements in the *New York Times* and the *Wall Street Journal*, which stated that 84% of New York-to-Chicago business travelers prefer Midway to American, TWA, or United Airlines. The advertisement apparently claims that 84% is a reasonable estimate of the percentage "of New York-to-Chicago business travelers who prefer Midway to American, TWA, or United Airlines." Here are three criticisms of this claim:

Criticism I – The ad isn't fair because USAir operates between New York and Chicago and wasn't included in the questions asked by this survey.

Criticism II – The ad isn't fair because most of the passengers neither filled out the survey nor returned it to the flight attendant.

Criticism III – The ad isn't fair because only Midway passengers were surveyed.

Which of these criticisms is (are) referring to a possible *sampling error,* and which of these criticisms is (are) referring to a possible *non-sampling error*? Explain briefly.

Online Problems for Chapters 1 - 4

16. *Professional Football Players:* www.nfl.com. The Web site of the National Football League (NFL) provides detailed information on players for each of the 32 teams in the league. Open the Web site www.nfl.com and locate the rosters for these teams. Open the roster for the Saint Louis Rams. Examine the data presented there. This roster presents a data set in which the individuals are football players.

   a. What variables are included in this data set?

   b. List two variables that are numeric in this data set.

   c. Give an example of a variable from this data set that would be summarized with a percentage. Explain.

17. *Surveys of Internet Users:* www.pewinternet.org. The Internet has become a big part of American life. To find out more about how Americans use the Internet, the Pew Internet & American Life Project conducts surveys and produces reports on Internet usage. Open the Web page www.pewinternet.org. Go to the "Reports" section of this Web site and choose a report from the public policy category. Open the complete report (in pdf form).

   a. What is the sample size for the survey?

   b. What is the margin of error for the survey?

   c. How were the subjects in this study contacted?

   d. What possible sources of non-sampling error do you see in the report?

18. *College Freshman survey:* What is the typical college freshman like? How do college freshmen feel about political issues like gay marriage? An ongoing survey of college freshmen is collected by many institutions such as Appalachian State University. Go to the website www.sllresearch.appstate.edu and click on "Research Reports." Review the 2003 Freshman Survey.

   a. What is the population of interest in this survey?

   b. Is this study a census? Explain why or why not.

   c. The report mentioned some surveys were not used because the subjects "seemed to not be answering questions honestly." Would this be a sampling or non-sampling error?

19. *Question wording:* Can question wording influence the results of an opinion poll? The Web site www.pollingreport.com provides the results of many current polls, including the questions asked, the sample size, margin of error, and several other features. Open this Web page and examine some of the polls they present.

a. Some of these polls present two separate versions of their questions to subjects to examine the effect of question wording. Page through the site until you find an example of one of these. (Typically, they will be marked as Form 1 and Form 2.) What was the exact question wording of these two versions?

b. Did the wording of these questions make a difference in the responses? If so, by how much? What aspect of the question do you feel was the cause of the difference?

Additional Problems for Chapters 5-6

20. In order to investigate a rumor that there is a greater than expected number of girls among the children of chemists, *Science* magazine conducted an informal survey of eight chemistry departments in 1992. A secretary in the chemistry department at Indiana University, Bloomington, thought there might be something to this rumor and made sure that every one of the 34 faculty members in her department who have children responded to the *Science* survey. Altogether, these Indiana chemists have 53 (56%) girls and 41 (44%) boys. Is this evidence that chemists produce more girl babies than expected? What kind of data would you prefer before drawing this type of conclusion?

21. A nutritionist is interested in whether excessive sugar intake causes hyperactivity in children. Two different study designs are contemplated to examine this issue.

    Design I – One hundred children with their parents' permission are to be subjects in the study. The nutritionist will randomly divide the children into two groups of fifty. Both groups of children will eat similar diets for a week, including a bowl of cereal, a sandwich, a main course, four glasses of a special milkshake, and two candy bars per day. However, in preparing the food in one of the groups' diets, all of the sugar is to be replaced with an artificial sweetener (so the children cannot tell the difference between the two diets). A psychologist who does not know which diet the children were on will then watch them at play and classify each child as showing signs of hyperactivity or showing no signs of hyperactivity.

    Design II – One hundred children will be watched while they play by a psychologist who will classify each child as showing signs of hyperactivity or showing no signs of hyperactivity. This data will then be compared with the results of a diet questionnaire filled out by the children's parents that gives information about the children's sugar intake over the previous week.

    Which design is an observational study and which is a comparative experiment? Which of these study designs is likely to be easier to carry out? Which design would give more convincing evidence about the relationship between sugar intake and hyperactivity? Explain.

22. A researcher at The Ohio State University believes that a certain component of ant venom can be used to lessen the amount of swelling in the knuckles of people suffering from arthritis. The ant venom treatment has been made into a capsule form that can be swallowed. Explain how you would design an experiment to investigate whether this new treatment, when taken orally each day for one week, causes a lower degree of swelling in arthritis sufferers. (You may suppose that 200 people suffering from arthritis have already volunteered to be experimental subjects.) Identify the explanatory variable and the response variable in your experiment.

23. The results of a recent survey showed that public speaking was the highest-ranking fear (for example, ranking higher than the fear of nuclear war). A researcher wishes to investigate the physical effects of public speaking. She wants to design an experiment to learn whether a typical person's blood pressure increases when he or she speaks in public. A blood pressure measurement is taken on 100 volunteers at the researcher's clinic. Fifty of these subjects are then selected and told to return the following day to give a 10-minute speech about their favorite restaurant in front of an audience of people who work at the clinic. A second blood pressure measurement is taken following the individual's speech. The other 50 subjects also return the following day to take a second blood pressure measurement.

   a. The researcher needs to decide between two strategies for choosing the 50 people who will be asked to give a speech:

      Strategy I – Put the names of the 100 subjects in a hat and draw out the 50 names of the people to be asked to speak.

      Strategy II – Ask for the first 50 volunteers from the group of 100.

      Which strategy do you prefer? Explain briefly.

   b. Could this be run as a double-blind experiment?

   c. Suppose that 1000 volunteers were used in the experiment instead of 100 (putting 500 in each group). How would that change the nature of the results?

24. A recent study at University of Maryland examined whether laughter is good for blood flow – which is presumed to be good for overall cardiovascular health. In this study, 20 healthy volunteers were randomly assigned to watch a comedy (choice of *There's Something about Mary*, *Kingpin*, or excerpts from *Saturday Night Live*) or the war drama *Saving Private Ryan*. Two days later, the same subjects came back and watched the type of movie they had not previously seen (for example, if they had seen *Saving Private Ryan* the first day, they would watch a comedy two days later). It turned out that 95% of the subjects had an increase in blood flow while watching the comedy and 75% of the subjects had a decreased blood flow while watching the war drama. A critic of the study mentions that it was not fair that the subjects were given a choice of comedies to watch but not given a choice of different war dramas.

   a. This study is:
      i) an observational study without a control group.
      ii) an observational study with the type of movie as a confounding factor.
      iii) an experiment.
      iv) not an observational study or an experiment.

   b. Whether or not subjects were given a choice:
      i) is an example of a confounding factor in this study.
      ii) is an example of blinding in this study.
      iii) is an example of a placebo in this study.
      iv) is an example of a control group in this study.

25. The article below appeared in the health section of the Web site CNN.com on July 14, 1999.

Affordable drug reduces mother-to-child HIV transmission, study says

July 14, 1999
Web posted at: 12:00 p.m. EDT

WASHINGTON (CNN) – An AIDS drug already available could prevent nearly 1,000 babies a day from contracting HIV at a cost developing countries can afford, according to a joint Uganda-U.S. study released Wednesday.

The study conducted in Uganda, which has one of the highest rates of HIV infection in the world, found two doses of the drug nevirapine can dramatically reduce the transmission of HIV from mother to child.

Over 600 HIV-infected mothers in Uganda participated in the study. Some 300 women were given a dose of nevirapine during labor, and their newborns were given another dose within three days of birth. The others in the trial were given the more traditional treatment of a short course of AZT.

Of more than 300 who took nevirapine, only 40 babies were infected with the HIV virus, while 77 of the newborns with the AZT treatment were infected.

---

a. What is the explanatory variable in this study? What is the response variable?

b. The article is short and does not give many details of the research. What would you need to know to determine if the research was done with a double-blind design?

26. A recent study showed that teenagers who had night lights in their rooms as babies are more likely to suffer from myopia (near sightedness) than teenagers who had not used night lights. But genetic factors are known to play a role, and near-sighted parents are more likely to install a night light in their children's room. This makes it hard to tell if the night light caused the subsequent myopia. This is an example of:
     i)   the double-blind technique.
     ii) internal inconsistency.
     iii) the placebo effect.
     iv) confounding.

Pick one and explain.

27. In 2003, a group of 100 employees of a large company were asked if they had a cold causing them to miss at least one day of work over the previous winter (i.e., the winter of 2002–2003). Sixty of these people did miss work due to a cold that winter. Starting in December of 2003, all 100 workers agreed to take a 1-gram tablet of vitamin C each day as a supplement to their regular diet. It turned out that only 50 of these employees missed at least one day of work over the winter of 2003–2004. The researcher conducting the study claims that this shows that vitamin C helps prevent colds and that the company would save money by providing free vitamin C tablets to all of its employees.

    a. Identify the explanatory variable and the response variable for this study.

    b. Describe a better experimental design that could have been used in an experiment starting in December 2003 using the 100 workers as subjects to find out whether vitamin C helps prevent colds. Explain why your experimental design is better than the one used (i.e., explain why it could provide better evidence of causation).

28. In a recent report, researchers at the Mayo Clinic examined tumor tissue from 105 patients with brain tumors. They found that the tumors that exhibited a specific change in the genetic material of chromosome one survived longer than patients whose tumors did not have this quality. They also pointed out that these patients tended to be younger than other brain tumor patients. This is an example of:
        i) the double-blind technique.
        ii) internal inconsistency.
        iii) the placebo effect.
        iv) confounding.

Explain your choice.

29. A doctor at a veterans hospital examines all of the patient records from 1985 to 2005 and finds that twice as many men as women fell out of their hospital beds during their stay. This is put forward as evidence that men are clumsier than women. Discuss briefly, including:

    a. whether this is an experiment or an observational study (and why).

    b. Whether this is an example of a possible confounding factor.

    c. a clear explanation of *why* it is a possible confounding factor.

30. Research published in the April, 2004 issue of *Pediatrics* followed the television viewing habits of 1278 children from the time they were one year old until they were seven years old. It turned out that one year-olds who watched three-to-four hours of television per day had a 30 to 40 percent increased risk of having attention problems when they are seven years old when compared with children who watched no television as one year-olds.

a. What is the explanatory variable? The response variable?

b. This is an example of:
   i) an observational study without a control group.
   ii) an observational study with a control group.
   iii) an experiment that was not randomized.
   iv) a randomized experiment.

Explain your choice.

31. Does the cocoa butter in chocolate raise serum cholesterol levels? Explain how you would design an experiment to answer this question. You may assume that 200 volunteers have agreed to digest pills containing cocoa butter on a daily basis for two weeks.

32. Researchers at Tufts University recently investigated whether calcium and vitamin D supplements can help the elderly avoid broken bones. Their research, published on September 4, 1997, in the *New England Journal of Medicine,* looked at 389 subjects aged 65 or older. The subjects were randomly divided into two groups. One group took pills containing 500 milligrams of calcium and 700 international units of vitamin D each day. The second group took pills each day that looked and tasted the same but contained only inert ingredients. Over the next three years it turned out that 6% of the group taking the calcium and vitamin D supplement had broken bones, whereas 13% of the people in the second group experienced a broken bone. Write a sentence describing the *design* of this research using the *three* most appropriate items from the list of 12 items below.

i) blinded subjects   ii) confidence   iii) reliability   iv) validity
v) comparative experiment vi) margin of error vii) sampling error   viii) parameter
ix) observational study   x) confounding   xi) statistical significance xii) randomized

33. In February 2000, the *Journal of the American Medical Association* published a report from a group of British researchers that babies born smaller than normal tend to have lower incomes as adults. The study tracked about 1000 full-term babies who weighed less than 2.5 kilograms at birth in Britain in 1970 and a control group of about 1000 babies born that year who had normal birth weights. At age 26 (i.e., in 1996) the subjects were interviewed and it was found that the adults who had been born small earned about 10% less than those who had been born at a normal weight. A critic of the study pointed out that low-birth-weight babies are more common among poor women. The critic has identified:

    i) a double-blind feature of the study.

    ii) bias caused by nonadherers.

    iii) an example of the placebo effect.

    iv) a lurking variable.

Pick one and explain.

34. A 1983 report in the *Journal of Ultrasound Medicine* described a laboratory investigation in which pregnant mice were randomly divided into two groups. One group received high doses of ultrasound while the other group did not. It turned out that the birth weights of the infant mice born to the ultrasound group were, on average, lower than the birth weights of the infants born to the mice who did not receive the ultrasound. This study created a worry that ultrasound might also reduce the birth weight of human infants. Researchers at the Johns Hopkins Hospital examined their hospital records and found 1598 infants who were exposed to ultrasound during their mother's pregnancy (i.e., the mother had a sonogram) and 944 infants who had not been exposed. In a 1988 article, they reported that the babies exposed to the ultrasound weighed less on average than the unexposed babies. They also pointed out that ultrasound is used for diagnostic purposes, often in pregnancies that are high risk for low birth weight. Thus, the difference between the birth weights of the babies exposed to ultrasound and the control group of babies that were not exposed may have been caused by the underlying medical reason for having the sonogram.

Which of the possibilities i through ix match the descriptions a through d. Explain.

a. The investigation reported in the *Journal of Ultrasound Medicine*
b. The Johns Hopkins investigation
c. The underlying medical reason for having the sonogram in the Hopkins investigation
d. The method used to create the comparison groups in the mouse investigation

i) an observational study	ii) a placebo	iii) the margin of error
iv) a nonsampling error	v) a lurking variable	vi) the judgment of experts
vii) randomization	viii) a comparative experiment	ix) double-blinding

35. A recent investigation examined study skills among college freshmen. The study randomly selected 200 students who had attended a study skills course and 200 students from the general university population who had not attended the course. The researchers found that students who had attended the course reported an average of 12.2 hours of study per week while the students who had not attended the course reported an average of 13.5 hours of study per week. The researchers were quoted as saying "The difference between these two groups is small but statistically significant." Explain in language that a person who has not taken a statistics course would understand what the researcher means by "statistically significant."

36. Does the herb Echinacea help the common cold? A study by researchers at the University of Wisconsin-Madison published a study in the *Annals of Internal Medicine* that found that Echinacea was no better than a placebo. The researchers randomly assigned 142 college students who recently came down with colds to receive either Echinacea in capsule form or a placebo in capsule form. The students did not know if they were getting the herb or the placebo and took their treatment for 10 days. The researchers reported that there was no statistically significant difference in the duration of the cold between these groups.

   a. What are the explanatory and response variables in this study?

   b. Was this study an experiment or an observational study?

   c. This study used a placebo. Explain why this would be needed in this study.

   d. This study showed that there was no statistically significant difference between the groups. Explain what this means in your own words.

37. Consider the following news story:

Study: Test of Nicotine Nasal spray. Can it help with long term smoking cessation?
Scientists in Reykjavik Iceland have completed a randomized controlled study to test a nicotine based nasal spray. Scientists recruited long time smokers to see if the nasal spray in addition to skin patches can help smokers kick the smoking habit. This study randomly assigned 237 to take a nicotine based nasal spray or a placebo in addition to using nicotine patches. The subjects were followed for 6 years with regular checkups and smoking cessation counseling.

---

   a. What is the explanatory variable in this study?

   b. Is this an experiment or an observational study? Explain.

38. State whether each statement is true or false. If false, what change makes it true?

   a. Using randomization to assign subjects to treatment groups helps to decrease bias in the interpretation of the results.

   b. The purpose of a placebo is to reduce the variability of experimental data.

39. Some researchers believe that vitamin E helps to prevent heart attacks. They point out that people who take vitamin supplements that contain vitamin E have a lower risk of heart attacks than people who don't take the supplements.

   a. Explain why the information cited by the researchers is not strong evidence that vitamin E lowers the risk of heart attacks.

   b. Describe how you would conduct a study to obtain the strongest evidence possible about whether vitamin E really helps to prevent heart attacks. You may assume that 200 patients with a high risk of having a heart attack in the next year are available to be subjects in your study.

40. A recent study suggested that women who work nights may increase their risk of breast cancer. The researchers theorized that bright light in the dark hours decreases melatonin secretion and increases estrogen levels. This study was reported in the *Journal of the National Cancer Institute* and was based on a sample of over 78,000 nurses. In reviewing the medical and work histories of these nurses, the researchers found that nurses working rotating night shifts at least three times a week were about 8% more likely to develop breast cancer. One of the researchers, Francine Laden, said "The numbers in our study are small, but they are statistically significant."

   a. What is the explanatory variable in this study? What is the response variable?

   b. Is this an experiment or an observational study? Explain.

   c. The researcher said that the results were statistically significant. Does this mean the results show a strong effect? Explain.

 EESEE Exercises

The following exercises make use of stories in the Electronic Encyclopedia of Statistics Examples and Exercises, or EESEE (pronounced ee-zee). EESEE is included in the E-STAT Pack that accompanies this workbook. You can also access EESEE at www.whfreeman.com/eesee.

41. **EESEE Story** *Sleeping Patterns in Ducks.* A recent report in the journal *Nature* examined whether ducks keep an eye out for predators while they sleep. The researchers, from Indiana State University, put four ducks in each of four plastic boxes arranged in a row. Ducks in the two end boxes slept with one eye open 31.8% of the time compared to only 12.4% of the time for the ducks in the two center boxes. Furthermore, the ducks in the center boxes did not keep one eye open more than the other, whereas the ducks in the outside boxes kept the eye facing away from the group open 86.2% of the time that only one eye was open. Is this an example of an observational study or a comparative experiment? Explain.

42. **EESEE Story** *Mud-wrestling* Could mud-wrestling be the cause of a rash contracted by University of Washington students in the Spring of 1992? Two physicians at the University of Washington Student Health Center sent a questionnaire to the 153 students who lived in a residence hall that had sponsored a mud-wrestling event about two weeks earlier. It turned out that 76% of the students who went in the mud reported having a subsequent rash, whereas only 4% of those that didn't go in the mud reported a rash during the same two-week period. This is an example of:

     i)   an un-randomized double-blind comparative experiment.
     ii) a double-blind observational study.
     iii) a randomized double-blind comparative experiment.
     iv) an observational study.

Explain your choice.

43. **EESEE Story** *Is Caffeine Dependence Real?* Whether it is from a double latte, a soft drink, or just black coffee, many people start their day with a dose of caffeine. Some people feel that they must have that dose or they will have physical pain. Is this pain real? The story "Is Caffeine Dependence Real?" presents a study designed to answer that question. Open this story and read through the Protocol.

a. What are the explanatory and response variables in this study?

b. Was randomization used in this study? Explain.

c. A placebo was used in this study. Give a detailed explanation of why the placebo was needed in this context.

44. **EESEE Story** *Is Caffeine Dependence Real?* Consider again the story "Is Caffeine Dependence Real?".

a. Was this a double-blind study?

b. Researchers were concerned that there may be a problem with non-adherence and took a step to determine if this was a problem. Explain what non-adherence would be in this study and what measure described in the study helped them determine this.

45. **EESEE Story** *Is Caffeine Dependence Real?* Once again, consider the story "Is Caffeine Dependence Real?". Is this a completely randomized design or a matched pairs design? Explain.

Online Problems for Chapters 5 - 6

46. *Playing with Fire:* The Building and Fire Research Laboratory of the National Institute of Standards and Technology is charged with testing the fire safety characteristics of materials found in various residential and commercial uses. Go to their website at www.fire.nist.gov and click on a link to data from an actual fire testing situation.

   a. What were the outcome variables measured in the experiment you found?

   b. What were the explanatory variables?

   c. Did the research laboratory replicate the experiment? What information can be gained by doing so?

Additional Problems for Chapter 8

47. The age of a pine tree was independently measured five times using a new electronic probe inserted in the tree's trunk. The measured values were 43 years old, 40 years old, 45 years old, 44 years old, and 41 years old. Later, this same tree was cut down and, by counting the growth rings, it was determined that the tree was really 34 years old. Does this new electronic device for measuring the age of trees have a greater problem with bias or with reliability? Explain.

48. The November 17, 1994, issue of *The New England Medical Journal* reported on a study of the effects of hormone therapy on middle-aged women. About 950 women took part in the study; half were selected randomly to receive the hormone therapy and the other half was given a placebo. After about a year, blood tests were conducted on each subject by a lab technician who was unaware of which group (treatment or placebo) the blood samples originated from. In presenting the results of the experiment, the authors reported that the women in the treatment group had experienced a statistically significant increase in HDL (the so-called "good" cholesterol) and a statistically significant reduction in LDL (the so-called "bad" cholesterol) when compared with the control group.

    a. Describe what is meant by the words "statistically significant" in the paragraph above.

    b. A newspaper report said the experiment had shown that hormone therapy is effective in reducing the women's risk of heart attack. This is not a justified conclusion because:
      i.) the design of the experiment was seriously flawed.
      ii.) the cholesterol measurements were not a reliable measure of the risk of heart attack.
      iii.) the cholesterol measurements may not be a valid measure of the risk of heart attack.
      iv.) there is no reason. The conclusion is justified.

      Pick one and explain briefly.

49. A teacher wishes to know if a new English curriculum will increase the creativity of her students in writing poetry. At the end of the unit, a sample of each student's poetry is given to a panel of five experts who rate the creativity of the poem on a 10-point scale. For each of the two statements below, tell whether it addresses the issue of measurement reliability or measurement validity.

    Statement I – In reporting the results, the teacher claimed that the experts' ratings demonstrated her students' creativity. A critic of the new curriculum argued that the ratings were a poor way to measure creativity.

    Statement II – The five experts seemed to rate each student nearly the same. For example, one student's poem got ratings of 8, 7, 8, 8, and 6 by the five experts.

50. A reporter filing a story about a new convention center reports that the largest meeting room at the center is about 50 yards long and about 20 yards across. These measurements are based on his pacing off the room in each direction and reporting the number of paces he took. In making this report, he faces two obvious criticisms. Of course, pacing off a room is inaccurate because of the inconsistency in a person's stride. If he paced off the room a second time, he would almost surely get a different answer. Also, reporting the answer in yards is inaccurate because the reporter is rather short and his paces are not typically a full yard long. Discuss the key statistical properties of the reporter's measurements.

51. A common measurement of the lifetime smoking habits of subjects in observational studies dealing with heart disease is the number called "Pack Years." This number is computed as:

(lifetime average number of packs smoked per day) x *(number of years smoked)*

A critic gives three reasons why "Pack Years" is not a perfect measurement:

Reason I – Often, when asked multiple times, subjects will give you different answers, about how much they smoked at various times in their lives.

Reason II – The average number of cigarettes smoked per day is not the only facet of a subject's smoking habits that is of interest. For example, some researchers believe that it is worse to smoke half pack a day for a year followed by a pack and a half a day for another year than to smoke a pack a day for two years.

Reason III – Subjects often lie about how much they have smoked, almost always tending to downplay the true amount.

a. Which of these criticisms argue about the validity of the Pack Years number?

b. Which one argues about the reliability?

c. Which one argues about bias in the measurement?

Explain your choices.

52. The diameter of the moon is independently measured four times by a process that is free of bias. The individual measurements come out 2157, 2166, 2162, and 2155 miles, which average out to 2160 miles. One more measurement is about to be taken using the same process. When compared with the estimate of 2160 miles, you would expect this next measurement to be:
   i.) more accurate as a measure of the true diameter of the moon.
   ii.) just as accurate at measuring the true diameter of the moon.
   iii.) less accurate as a measure of the true diameter of the moon.

Pick one and explain.

53. Trucks are weighed at a truck scale to establish the amount owed in road taxes. Someone complains that the weighing procedure has three problems.

> **Problem I** – Sometimes the driver is sitting in the truck when it is weighed.

> **Problem II** – When the same truck is independently weighed more than once, the truck scale will give different values.

> **Problem III** – When the legislature established the road tax, they intended to tax according to the value of the goods being shipped, not according to the weight.

> Which of the above indicates:
> a. a problem with bias?
>
> b. a problem with reliability?
>
> c. a problem with validity?

> Explain your choices.

54. Apgar scores are a measurement of an infant's overall health taken a few minutes after birth. The score ranges from 0 (dead) to 10 (perfect health) and are based on tests of the baby's heart and breathing rate, muscle tone, etc... A critic gives three reasons why the Apgar score isn't a perfect measurement:

> **Reason I** – There are many important facets of health that aren't measured by the score.

> **Reason II** – A doctor's rating may be affected by being present at the birth, often giving unwarranted low values to babies whose birth was difficult.

> **Reason III** – Two different doctors may give different Apgar scores, even when measuring the same baby at the same time.

> a. Which of these criticisms argue about the validity of the Apgar score?
>
> b. Which one argues about the reliability?
>
> c. Which one argues about bias in the measurement?
>
> d. Suppose two doctors both judge an infant's health using the Apgar system and the average of their two values is taken as the "official" Apgar score. How will this impact the validity, reliability, and bias of the measurement? Explain.

55. A team of engineers develop a new device to measure the speed of lava flows in volcanic areas. The device uses an infrared laser to measure the speed. The researchers set up a test in Hawaii to measure a lava flow that is moving at a known speed of 20 feet per minute. In a test of 10 samples, the engineers find the following measures: 25, 12, 19, 20, 18, 22, 28, 15, 21, and 20. Does this device have a bigger problem with bias or reliability? Explain how you know.

56. Two investigators measure the depth of an oceanic trench. The first one makes 20 independent measurements and reports the average of the values obtained; the second person reports the result of a single measurement using the same process.

   a. Which investigator will have more reliability in the number reported, or will they both have about the same reliability?

   b. Which investigator will have more bias in the number reported, or will they both have about the same bias?

57. Two chemists (Joe and Bob) make measurements of the amount of lead in a sample of the city's water. The chemists each repeat their measurements 20 times. In reviewing the results, the public works manager finds that the measurements made by Joe had a much higher standard deviation than the measures made by Bob, but the averages for Joe and Bob were the same. This implies:
   i) Joe has made more reliable measures than Bob.
   ii) Joe has made more biased measures than Bob.
   iii) Joe has made less reliable measures than Bob.
   iv) Both ii and iii.

58. A nutritionist for the FDA develops a new machine that quickly measures the amount of fat in foods. The nutritionist's supervisor examines this machine and has the following comments:

   "When I tested the machine on a standard sample that we know to have 25 mg of fat we found it always gave values that were too high, although they only varied by plus or minus 1 mg. Also, the machine measures caloric content rather than true fat content."

   Discuss the statistical measurement properties of the new machine.

Additional Problems for Chapters 10 - 12

59. Consider the following pictogram.

> **An Apple A Day Keeps Costs Away**
>
> Many corporations are implementing wellness programs for employees. These programs keep costs of absenteeism and health care down. A survey of major corporations revealed the following percentage of surveyed firms that participate in each type of programs.
>
> **Wellness Information**
> 🍎🍎🍎🍎🍎🍎 56%
>
> **Cholesterol screenings**
> 🍎🍎🍎🍎🍎 48%
>
> **Smoking Cessation**
> 🍎🍎🍎🍎 31%

a. Explain why the percentages presented in this pictogram could not be represented as a pie chart.

b. A graphic artist suggests an alternative pictogram pictured below. Which pictogram, the top or the bottom, gives the more honest representation of the survey percentages? Explain.

> **An Apple A Day Keeps Costs Away**
>
> Many corporations are implementing wellness programs for employees. These programs keep costs of absenteeism and health care down. A survey of major corporations revealed the following percentage of surveyed firms that participate in each type of programs.
>
> **Wellness Information**
>
>
>
> 56%
>
> **Cholesterol screenings**
>
>
>
> 48%
>
> **Smoking Cessation**
>
>
>
> 31%

60. An economist produced this graphic with an article stating that gas prices did not move much between November 2004 and July 2005. Explain why the graph does not give a fair picture of the changes in gas prices in this period.

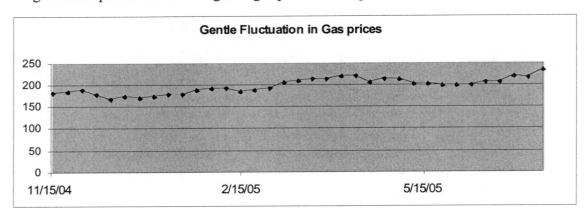

61. Explain briefly what's wrong with the graphical presentation below (aside from the depiction of George Washington):

THE SHRINKING BUYING POWER OF THE DOLLAR

1970: $1.00

1980: $0.47

1990: $0.29

2000: $0.22

62. This line plot shows the average national price for 500 KWH of electricity (in dollars). Explain how this graphic illustrates seasonal variation. Why would you expect to see seasonal variation of this type? Explain.

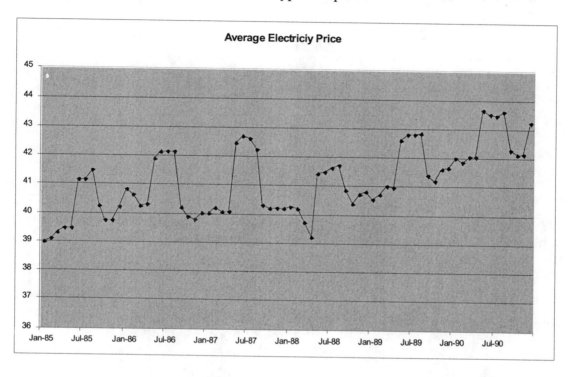

Average Electriciy Price

63. Which of the following is most likely to have a distribution that is skewed to the right? Pick one and explain.
    i)   The age of students in a third grade class.
    ii) The height of male soldiers in the United States Army.
    iii) The cost of homes in the Miami Florida.

64. A college professor recorded the year in school for the students in his introductory history class. This bar chart was produced from the resulting data. One of the pie charts below was also created using these data. Which of these pie charts was created with these data? Explain how you know.

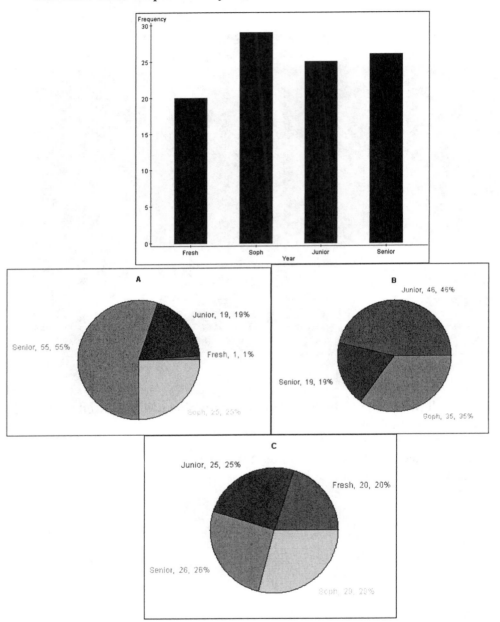

65. The ages of U.S. presidents at the time of their inauguration were used to create this stem and leaf plot.

```
Stem-and-leaf of Age        N  = 43
Leaf Unit = 1.0
   4 23
   4 667899
   5 0111112244444
   5 555566677778
   6 0111144
   6 589
```

    a. Teddy Roosevelt was the youngest person inaugurated president. How old was he?

    b. Ronald Regan was the oldest person inaugurated president. How old was he?

    c. How many presidents were 55 years of age when inaugurated?

    d. Which of the following best summarizes these data?
       i) Most presidents are over 60 when inaugurated.
       ii) Most presidents are in their 50s when inaugurated.
       iii) There have been several presidents who were in their 30s when inaugurated.
       iv) All presidents were over 50 when inaugurated.

66. A large movie studio was interested in the audience that was attending a movie. A random sample of six ticketed moviegoers was collected. The ages of these individuals were as follows: 17, 16, 21, 32, 14, and 41. Describe the center and spread of these data using statistical summaries.

67. Most of the students in last year's statistics course that responded to a survey did not smoke at all, although one student smoked two packs a day. True or false: For this group of students, the mean number of cigarettes smoked per day was larger than the median.

68. The school taxes in a small town average $1200 per household with a standard deviation of $600. Next year, the school taxes of every household in the community will be raised by $60 to cover the cost of a new high school gymnasium. The new tax bills will then average $_____ with a standard deviation of $_____. Fill in the blanks and explain briefly.

69. The workers at a small company have a mean salary of $28,000 per year with a standard deviation of $8,100 per year. If every employee gets a $900 per year raise, the new salaries would have a mean of $_____ with a standard deviation of $_____ per year. If instead, every employee gets a 5% raise, then the new salaries would have a mean of $_____ with a standard deviation of $_____ dollars per year. Fill in the blanks and explain briefly.

70. An instructor of a large college class gives an exam that has a possible total of 100 points. The instructor records the scores of 100 students from his class and produces the following histogram. The instructor says any score above a 90 is an A. Scores in the 80 to 89 range would be a B, scores between 70 and 79 would yield a C, and below 70 would result in a failing grade.

a. How many individuals scored between 80 and 84?

b. Which of the following statements best summarizes this situation? Pick one and explain.
   i) Almost everyone received a failing grade.
   ii) Most students received a C on this exam.
   iii) Most students received an A on this exam.
   iv) More than 10 students received a C for the course.

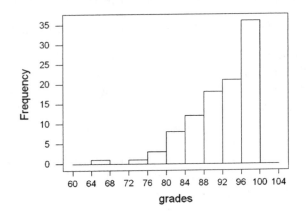

71. Match the summary statistics with the histograms. Explain your choices.

a. mean = 6.6, median = 6.8, standard deviation = 1.3 _____

b. mean = 6.6, median = 6.0, standard deviation = 8.65 _____

c. mean = 6.6, median = 3.75, standard deviation = 7.4 _____

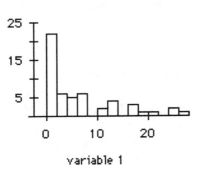

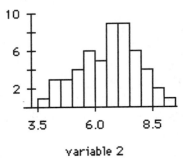

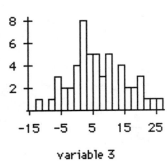

72. Consider the histogram below.

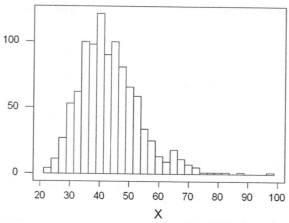

X

a. Would this histogram be considered skewed or symmetric? Explain.

b. This histogram is most likely to represent
   i) the ages (in years) of 1000 home owners in Illinois.
   ii) the height (in inches) of 1000 randomly selected 7 year-old children from California.
   iii) the end-of-semester grade percentage of 1000 students in an introductory statistics course.
   iv) the annual income (in dollars) for 1000 randomly selected home owners in Illinois.
   Pick one and explain.

73. Consider the following histograms:

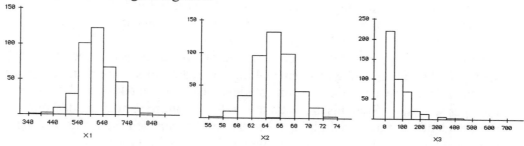

a. In which of these distributions would the mean and median be most different?

b. How would the standard deviation of variable X1 compare with standard deviation of variable X2? Explain.

c. The histograms above are the results of questions asked of a group of undergraduate students. Match the histogram above to the appropriate question below.
   i)   What is your mother's height (in inches)?
   ii) How many minutes did you watch television yesterday?
   iii) What was your SAT verbal score?

74. A survey of students in an introductory statistics course at Pennsylvania State University recorded the cumulative college GPA (on a 4-point scale) of 210 individuals. The resulting data was used to produce the following boxplot.

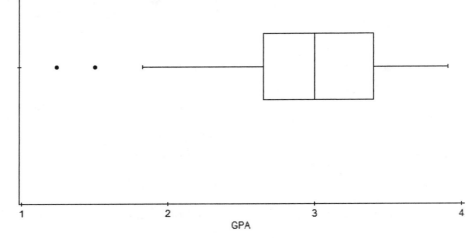

GPA

   a. What is the median GPA of these students?

   b. What is the third quartile of this distribution?

   c. Would this distribution be best described as skewed left, skewed right, or symmetric? Explain

75. The customers who purchase items from a particular mail-order catalogue spend an average of $180 with a standard deviation of $100 on the products they order. The mail order company also charges every customer an additional $5 for shipping and handling. What is the average and standard deviation of the final amount billed (i.e., cost of products + shipping and handling)? The average bill is $_____ with a standard deviation of $_____. Explain how you came to your answers.

76. A large university records the number of credit hours taken during the semester by its students. At that university, students taking 13 credit hours or more are considered full time, but there are some part time students that may be taking as few as 1 credit hour per semester. Students are not allowed to take more than 20 credit hours, but most students take at least 14 hours. For the most recent semester, the university produced the following summary for the number of credit hours.

Mean	Median	Min	Max	Q1	Q3	St Dev.
15.5	16	1	20	14	17	2.1

   a. Would these data be considered to be skewed left, skewed right, or symmetric? Explain how you know.

   b. Students at this university pay $550 tuition for each credit hour. What is the standard deviation for the amount students pay in tuition. Explain fully and show your calculations.

77. Fortune magazine annually publishes a list of the wealthiest billionaires. In 1992, there were 80 billionaires living in Europe. The wealth (in billions of dollars) of these 80 individuals was used to create the following boxplot.

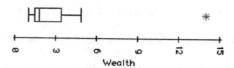

a. The median wealth for these 80 billionaires was around _____.

b. The wealthiest 25% of these individuals have over _____ billion dollars.

c. Queen Elizabeth II of England was the wealthiest of this group and appears as an outlier. Her wealth is around _____.

d. Is the distribution of wealth of these billionaires skewed or symmetric? If it is skewed, to what direction is it skewed? Explain how you know.

78. Match the summary statistics with the histograms. Explain your choices.

a. mean = 4.99, median = 3.13, standard deviation = 5.49 _____

b. mean = 4.89, median = 4.83, standard deviation = 7.99 _____

c. mean = 5.01, median = 6.87, standard deviation = 5.49 _____

d. mean = 4.96, median = 4.93, standard deviation = 0.96 _____

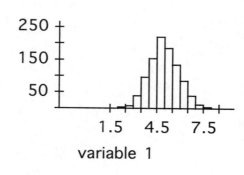

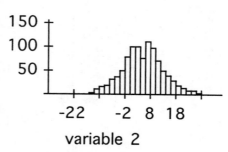

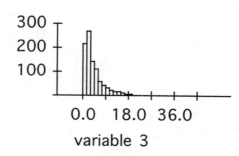

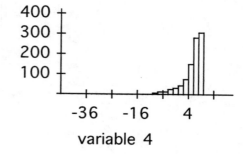

91

79. During the fall term in 1997, students in an introductory statistics course at the Ohio State University completed a survey in which they reported, among other things, their height (in inches) and gender. Using the results of the 474 completed surveys, separate boxplots for females (f) and males (m) were created and are given below.

a. What is the median height for males in this group of students?

b. What is the third quartile for females in this group of students?

c. The middle 50% of male students were between _____ and ____ inches tall.

d. The heights of males would best be described as:
   i) skewed left
   ii) skewed right
   iii) symmetric

Pick one and explain how you know.

e. Is the height of females bimodal?
   i) Yes.
   ii) No.
   iii) It is impossible to tell from this plot.

f. Is the following statement true? "In this class, over 25% of female students are shorter than the shortest male student." Explain how you know.

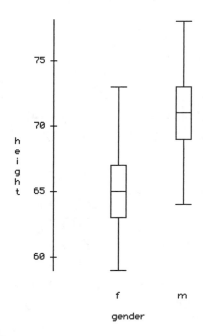

92

80. The 2004 presidential election was a very tight race between the two principal contenders, President George W. Bush and Senator John Kerry. The election was particularly close in the state of Ohio where many controversies resulted regarding problems with the election process, such as long waiting times at polling places. A survey of 473 Ohio students asked whether they voted for President Bush or for Senator Kerry and how long they waited to vote (students who voted at their polling place only). Boxplots of the data are shown below.

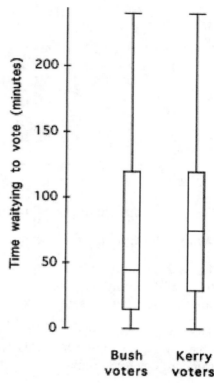

   a. The 75th percentile for the amount of time the Bush voters waited was:
      i)   higher than for Kerry voters.
      ii) lower than for Kerry voters.
      iii) about the same as for Kerry voters.
      iv) impossible to judge from the plot.

   b. The median amount of time the Bush voters waited was:
      i)   higher than for Kerry voters.
      ii) lower than for Kerry voters.
      iii) about the same as for Kerry voters.
      iv) impossible to judge from the plot.

   c. The distribution of the amount of time the Bush voters waited is:
      i)   somewhat skewed to the left.
      ii) somewhat skewed to the right.
      iii) perfectly symmetrical.
      iv) seen from the boxplot to be bimodal.

81. For a small class, the standard deviation of the scores on a test is 0. This says that:
      i)   The scores are all 0%.
      ii) The scores are all the same.
      iii) The scores have both negative and positive values.
      iv) The scores are all 100%.

82. A teacher of high school mathematics teaches a class of 50 students. The teacher gives two exams to the students. The scores on these exams are used to produce these boxplots.

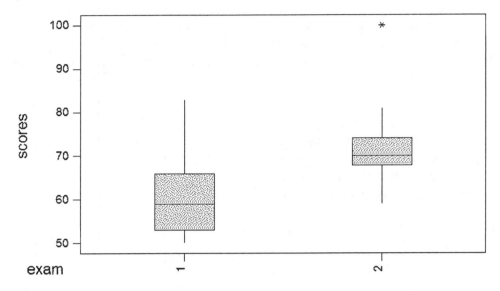

a. What is the median of exam 2 (approximately)?

b. What is the third quartile of exam 1 (approximately)?

c. There is an outlier in the scores for exam 2. The standard deviation for exam 2 is 6.03. If the outlier is removed, would the standard deviation of exam 2 be larger, smaller, or about the same? Explain.

d. The teacher would like to adjust the scores on exam 1 to be comparable to those on exam 2. To do this, she adds 10 points to every student score. Will the standard deviation of the scores for the exam change? Explain.

83. True or false: If the standard deviation of a list is zero, then all the numbers on the list are the same.

84. True or false: If the means and standard deviations of two different lists of numbers are the same, then all of the numbers on the two lists are the same.

85. True or false: A standard deviation cannot be negative.

86. For which of the three histograms below is the median larger than the standard deviation? Explain your choice.

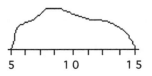 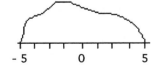 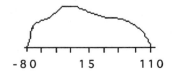

94

EESEE Exercises

87. **EESEE Story** *Health Care Graphics.* The problem of rising health-care costs has been a problem for many years. The story Health Care Graphics discusses an advertisement by a health-care group that presents how costs changed between 1986 and 1992. Open the story, read the introduction and examine the graphic. Is this a fair graphical representation of the increase in Emerald's claim costs over the years 1986–1992? Justify your answer with an explanation of what features of the graph you like or don't like.

88. **EESEE Story** *Hunting in the Rain Forest.* Researchers Dwyer and Minnegal studied the hunting capability of the residents of the Gwaimasi Village in the New Guinea rain forest. Open the story and read through the introduction. The Results tab of this story provides a table that includes variables on 15 males in a village in New Guinea.

    a. Create a stem and leaf plot for the weight of cassowary bird these males captured. How would you describe the shape of this distribution?

    b. We would like to summarize the typical amount of cassowary bird captured by these individuals. Would you use the mean or median? Explain.

Computer Exercises

Creation of graphics and calculating summary statistics is much easier using computer software. Use statistical software to complete the following exercises.

89. Open the datafile called "NFLRoster". These data gives the heights, weights, birthdates, and positions of players in the National Football League.

    a. Consider each of the following variables and give your guess and an explanation as to what type of shape you would expect for each of the histograms.
        i. Most football players go into the NFL right after college. They then typically stay in the NFL for around 5 years, although some stay for as many as 20 years. What shape would you expect for the year-of-birth variable?
        ii. The day of birth for all players?
        iii. The height of all players?

    b. Create a histogram of each of the variables. How do these histograms compare with your answers in the previous part?

    c. Create a histogram of the weight variable, using a bin width of 10 pounds. What shape does this histogram take on?

d. Create a boxplot of the weight of all players. Does this boxplot show all of the relevant features?

e. A football team consists of two types of players, linemen who are especially large and backs who specialize in handling the ball and running. Explain how these two groups are represented in the histogram you created in part c.

90. Open the datafile called "Au91survey" and find the variable 'books$' – the amount of money that the students in the survey reported spending for textbooks in fall 1991. Make a histogram of 'books$.'

a.   Try to guess the mean amount of money spent on textbooks and the standard deviation by looking at the histogram. What are your guesses? Use software to compute the exact values and record them.

b. Suppose each student in the class taking the survey has to buy an additional book costing $12.95. What should be the new mean amount spent on textbooks? What will the standard deviation be? Use software to verify your answers.

c. Obtain the histogram of the 'books$' variable and the histogram of the derived variable that is 'books$'+12.95 (part b). How does this histogram compare to the histogram of the original variable, 'books$'? Consider the shape of the histogram, the location, and the spread.

d. Suppose the bookstore's profit is one-tenth of the price paid for a textbook (multiply each value of 'books$' by 0.1). What is the bookstore's mean profit from each student? What is the standard deviation? Use software to verify your answers.

e. Obtain the histogram of the derived variable that is 'books$'*0.1 (part d). How does this histogram compare to the histogram of the original variable, 'books$'? Consider the shape of the histogram, the location, and the spread.

Online Problems for Chapters 10 - 12

Open the Web site www.shodor.org/interactivate/activities/boxplot this Web page presents a Java applet that produces boxplots of several data sets.

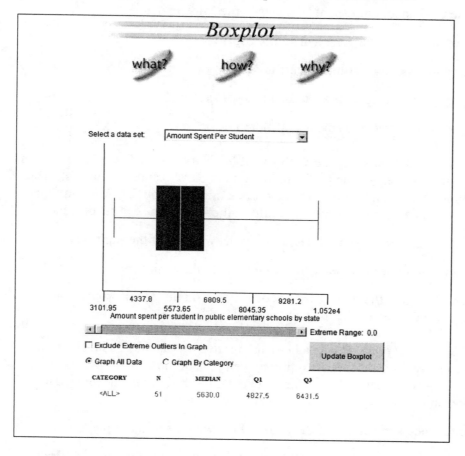

91. One of the data sets provided by this applet is the amount of money the 50 states and the District of Columbia spent per student on elementary schools. For example New Jersey spent $10,180 per student in 2000. Examine the boxplot of these data.

   a. What is the median amount spent?

   b. The top 25% of states spent over _____.

   c. What is the inter-quartile range for these data?

92. This data set also has information about what region of the country (North East, Midwest, South, or West) that the state is located. Click the radio button for **Graph By Category.** This will produce 4 different boxplots that correspond to the different regions. Examine these boxplots.

    a. What is the median amount spent by states in the Midwest?

    b. What region has the highest median amount spent?

    c. What region has the highest inter-quartile range?

    d. What region would be considered to be the most skewed?

    e. What region would have the smallest standard deviation? Explain.

93. *Professional Basketball players:* The Web site of the National Basketball Association at www.nba.com provides detailed information on the players and their teams. The 2005 champions of the National Basketball Association were the San Antonio Spurs. The page www.nba.com/spurs/roster gives information about the Spurs' players.

    a. Calculate the mean and standard deviation of the weights of these players.

    b. Calculate the 5-number-summary of the weights of these players.

    c. If you wanted to give the typical weight of a player on this team, would you choose to use the mean or the median? Explain.

94. *Professional Basketball players:.* Another team in the NBA is the Miami Heat, who are known for one of their most outstanding players, Shaquille "Shaq" O'Neal. Open the page www.nba.com/heat/roster. This page includes the heights, weights, dates of birth, and other information about the players of the Miami Heat.

    a. Calculate the mean and standard deviation of the weights of these players.

    b. Calculate the 5-number-summary of the weights of these players.

    c. Shaquille O'Neal is much heavier than other players on this team and would probably be considered an outlier. Remove him from the data set. Which changes more when he is removed, the average or the median? Does the standard deviation change?

Additional Problems for Chapters 13

95. The histogram below was created from the cumulative college grade point averages (GPA) of the students in a large statistics class. One student had a GPA of 2.37. The standard score for his GPA would be about:

     i)   -2.8

     i)   -1.0

     ii)   0.3

     iii) 1.3

Pick one and explain your reasoning.

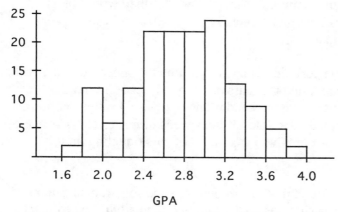

96. Bags of potatoes in a shipment averaged 10 pounds with a standard deviation of 0.5 pounds. A histogram of these weights followed the Normal curve quite closely.

   a. What percentage of the bags weighed less than 10.25 pounds?

   b. What percentage weighed between 9.5 and 10.25 pounds?

97. This is a histogram of the average rainfall (in inches per year) of the 60 largest cities in the United States. Minneapolis gets about 25 inches of rainfall per year. Its standard score in this data set would be:

   i) higher than 3.

   ii) lower than -3.

   iii) a negative number between 0 and -3.

   iv) a positive number between 0 and 3.

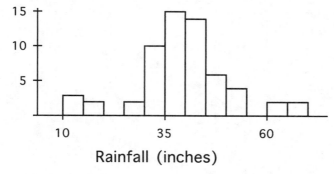

Rainfall (inches)

98. True or false: If a distribution is skewed to the left, then the median will typically have a negative standard score.

99. A citrus fruit company ships bags of oranges throughout the country. The weight of these bags of oranges averaged 8 pounds with a standard deviation of 0.5 pounds. A histogram of these weights followed the Normal distribution quite closely.

    a. What percentage of bags weigh less than 8.75 pounds?

    b. The shipping manager for this company has decided that they would like to pull aside the lightest 10% of these bags for repackaging. 10% of these bags are below what weight in pounds?

100. An engineer with the state transportation agency monitors the speed of cars traveling on a section of highway. He has found that cars passing a checkpoint have an average speed of 51 miles per hour, with a standard deviation of 3 miles per hour. The official has also found that the histogram of the speeds of these cars is very close to a normal distribution. How fast are the fastest 20% of cars traveling on this highway?

101. The owner of a restaurant has found that the amount of soda sold at his restaurant averages 75 liters per day. The owner has also found that the daily usage follows a normal distribution with a standard deviation of 7.2 liters.

    a. What percentage of days does the restaurant use more than 98 liters of soda?

    b. On only 10% of the days did the restaurant use less than _____ liters of soda. Fill in the blank and show your work.

    c. There are 0.26 gallons in a liter. If we had measured the amount of soda in gallons rather than liters, would the standard deviation of the amount of soda sold now become 7.2 gallons? Explain why or why not.

Online Problems for Chapter 13

Open the Web site: www.whfreeman.com/applet/normal.html.
This Web page has an applet that calculates normal probabilities. On the applet you will see the outline of a normal distribution. You will also see three entry boxes across the bottom of the applet, into which you can enter the mean and standard deviation of a particular normal distribution. You can also choose to have 2-tailed areas that are equally spaced on either side of the mean. The applet has flags that can be moved to choose different sections of the normal distribution. As you move the flags, the proportion of the distribution between the flags appears in a box.

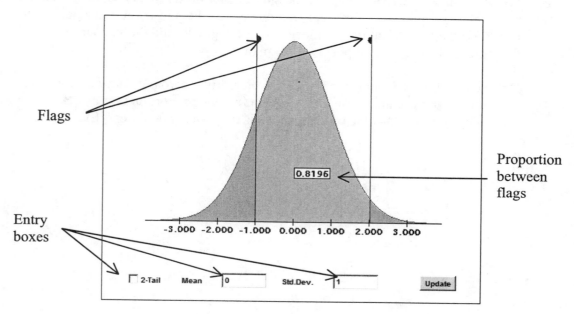

102. A college professor recorded the number of miles he drove in his car each week during the school year. He found that the distribution of this mileage was very close to a normal distribution and had an average of 236 miles with a standard deviation of 31 miles. We can use the normal distribution calculator to work with this distribution.

a. Enter the mean and standard deviation in the appropriate boxes in the applet and examine the scale across the bottom of the applet. What are the largest and smallest values we see in the distribution?

b. We would like to know what percentage of the time the professor travels between 200 and 300 miles per week. Insure that the 2-Tail entry box is unchecked and set the flags at the 200 and 300 mile points. Approximately, what proportion of weeks during the school year fall in this range?

c. We would like to know how often the professor drives over 200 miles in a week. Move one flag to the extreme right location on the graph and move the other flag to the 200 mile mark. What proportion of the values is between these markers?

d. What percentage of the time does the professor drive more than 300 miles? You can find this answer using the answers to the previous two questions. Look at the graphic and think about how to find the answer, then calculate your answer using your method. Next, move the flags to the extreme right of the graph and to 300 and confirm your answer.

e. We would like to determine what mileage has 15% of the weeks below it. Set one flag at the extreme left of the graph and move the other until you find a point that has 15% below it. What is the mileage at that flag? Would the standard score of this mileage be negative or positive? Explain.

f. During the summer months the Professor often drives across country on vacation. Would it be appropriate to use this applet to find the proportion of all weeks (including summer) that the professor travels more than 400 miles? Explain.

Additional Problems for Chapters 14 - 15

103. A scientist is studying the relationship between the depth of a watermelon vines' roots and the weight of the watermelons produced. The scientist collects measurements from a random sample of vines. She then conducts a test of hypothesis in which the null hypothesis is that there is no correlation between the two variables (correlation = 0) versus the alternative that the correlation is greater than 0. From this test she found a P-value of 0.0032 (0.32%). What does this tell us?

   i)   The correlation is significantly greater than 0.
   ii) The correlation is not significantly greater than 0.
   iii) The correlation is very small.
   iv) The correlation is 1.

Explain your choice.

104. True or false: If the slope of the line is 1, then the correlation must also be 1.

105. If the standard deviation of x is equal to the standard deviation of y then the slope of the regression line relating x and y will be:
   i)   equal to 1.
   ii) equal to the correlation.
   iii) equal to the mean of x.
   iv) equal to the standard deviation of x.

Explain your choice.

106. Consider the following two correlations:

   I – The correlation between weight (in pounds) and height (in inches) for all of the babies born in Franklin County this year.

   II – The correlation between weight (in kilograms) and height (in centimeters) for all of the babies born in Franklin County this year.

   There are about 2.2 pounds to the kilogram and about 2.54 centimeters to the inch. Which statement is true?
   i)   Correlation **I** is larger.
   ii) Correlation **II** is larger.
   iii) Correlations **I** and **II** are equal.
   iv) There is not enough information to tell which correlation is larger.

Explain your choice.

107. A realtor took a random sample of records of sales of homes from the files maintained by the Albuquerque Board of Realtors. From this sample he recorded the amount paid in real estate taxes (in dollars), and the sales price of the home (in thousands of dollars). From this information the following output was created.

```
Dependent variable is:              TAX
No Selector
117 total cases of which 10 are missing
R squared = 76.7%      R squared (adjusted) = 76.5%
s = 149.5 with 107 - 2 = 105 degrees of freedom

Source        Sum of Squares   df    Mean Square   F-ratio
Regression    7.71954e6         1    7.71954e6     345
Residual      2.34782e6        105   22360.2

Variable    Coefficient   s.e. of Coeff   t-ratio    prob
Constant    36.3444       43.24           0.841      0.4025
prices      7.02784       0.3782          18.6       ≤ 0.0001
```

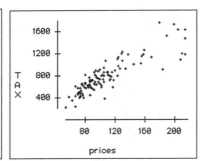

a. What is the correlation between the sales price and the taxes paid?

b. We know that a home sold for a price of $180,000. Use the least squares line presented above to predict the average taxes for homes that sold at this price.

c. We know that a home sold for $400,000. Would it be appropriate to predict the average taxes for a home that sold at this price? If so, make the prediction. If not explain why not.

108. Indicate which of the following statements are true and which are false? If a statement is false, explain why.

a. A correlation of 0.8 means that 80% of the points in the scatterplot lie above the regression line.

b. If the correlation between two lists of numbers is zero, then there can be no relationship between them.

c. For all of the books in the Library of Congress, the correlation between the thickness of the books (in inches) and their number of pages would be positive.

d. For all of the cars registered in the state of Ohio, the correlation between their fuel efficiency (in miles per gallon) and their weight (in pounds) would be positive.

e. If the correlation between height (in inches) and weight (in pounds) for a group of people is 0.7, then the correlation between their heights (in centimeters) and their weights (in kilograms) will still be 0.7.

109. A survey of homes in Whitehall, Ohio recorded the market value (market) of the home, the size of the home in square feet (sqft), and the number of bedrooms (bed) that each home had. Two resulting regression outputs are given below:

```
The regression equation is
market = - 34778 + 94.2 sqft

Predictor          Coef        StDev           T          P
Constant         -34778         4860       -7.16      0.000
sqft             94.201        3.507       26.86      0.000

S = 28294        R-Sq = 75.2%      R-Sq(adj) = 75.1%
```

```
The regression equation is
market = - 4580 + 29845 bed

Predictor          Coef        StDev           T          P
Constant          -4580        14032       -0.33      0.744
bed               29845         4479        6.66      0.000

S = 52150        R-Sq = 15.7%      R-Sq(adj) = 15.4%
```

a. What is the correlation between the number of bedrooms and the market value?

b. We know that a home in Whitehall has three bedrooms. What market value would we predict for this home?

c. If you can choose either square footage or number of bedrooms to use as a predictor of market value, which would make the best predictor? Explain.

110. True or false: If the correlation between two variables is negative, then high values of one variable tend to be associated with low values of the other variable.

111. True or false: When the correlation between two variables is nearly negative 1, then there is a cause-and-effect relationship between them.

112. A study measures the average annual snowfall (in inches) for 10 cities over the last decade along with the greatest Earth movement (on the Richter scale) over this same time period. The study included data from five cities in California's San Francisco Bay Area and five cities from Canada's province of Ontario. The study found a very strong negative correlation between the two variables. Does this mean that a strong snowfall will prevent earthquakes? Explain your answer briefly (identify the type of spurious argument involved and draw a picture to illustrate).

113. Over 400 students in a statistics class were asked their GPA in high school and their GPA so far in college. The results were analyzed giving the following regression output:

```
Dependent variable is:                    GPA (OSU)
No Selector
474 total cases of which 21 are missing
R squared = 23.1%      R squared (adjusted) = 22.9%
s = 0.4388  with  453 - 2 = 451  degrees of freedom

Source        Sum of Squares    df    Mean Square    F-ratio
Regression    26.0375           1     26.0375        135
Residual      86.8298           451   0.192527

Variable      Coefficient    s.e. of Coeff    t-ratio    prob
Constant      1.26652        0.1421           8.91       ≤ 0.0001
GPA (high …   0.501958       0.04316          11.6       ≤ 0.0001
```

One student in the class had a high school GPA of 3.2. What would you predict for her GPA at Ohio State? Show your work.

114. As the price of gasoline increases many people are considering purchasing gasoline electric hybrids. Are hybrids really different from other cars? We can use scatterplots and correlation to explore the relationship of variables and see how hybrids fall in these groups. Use the computer software of your choice to open the data set "Cars2006."

   a. Create a scatterplot between the highway and city gas mileage. Illuminate the hybrid cars like the Toyota Prius, Honda Insight, Toyota Highlander Hybrid, Lexus RX400H, and Ford Escape Hybrid. Do these cars appear as outliers in the scatterplot?

   b. If the Toyota Prius and Honda Insight were removed from the scatterplot, would the correlation increase or decrease? Explain.

   c. Create a scatterplot between the highway mileage and engine displacement. Do the hybrids appear as outliers in this scatterplot? Explain.

115. Below is a scatterplot of the relationship between the Infant Mortality Rate and the Percent of Juveniles Not Enrolled in School for each of the 50 states plus the District of Columbia. The correlation is 0.73. If the District of Columbia (identified by the X) had been left out of the data set, then the correlation between these two variables for the 50 states would:

      i)   be higher than 0.73.
      ii) not change at all.
      iv) be lower than 0.73.

Pick one and explain briefly.

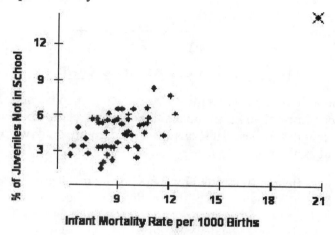

116. Open the data file called "79Cars." This data gives information about the 1979 model year cars taken from *Consumer Reports* magazine.

a. Examine the weight of the cars (notice that the weights are given in units of thousands of pounds). Make a histogram and describe its shape. Calculate the values needed for the five-number summary and draw the boxplot. Does the five-number summary do a good job of describing the weight distribution?

b. Suppose we wish to know how fuel efficiency (the variable called 'MPG' gives miles per gallon) is affected by the weight of the car. Make a scatterplot of the data. Is there a positive or negative association between these two variables? What is the correlation? Does the association look linear? Explain why you chose one variable to be *x* and the other to be *y*.

c. Instead of *Miles per Gallon*, change the fuel efficiency variable to *Gallons per Mile* (GPM = 1/MPG). Make a scatterplot of GPM versus the weight of the car. How does this scatterplot differ from the one you made in part b?

d. What is the correlation between the weight and the GPM of the cars? We want to predict the fuel efficiency for a 1979 model year car that weighs 3000 pounds. Would it be appropriate to use the regression method to make this prediction? If yes, what is the prediction? If no, explain why not.

117. The height, in cm, and length of the middle metacarpal bone, in mm, of 10 skeletons were measured. (The metacarpal bones are in the hand between the wrist and fingers.) The scatter diagram is given below.

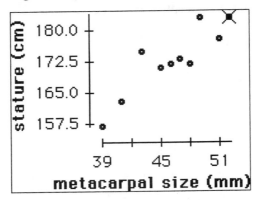

a. If the height and metacarpal length of the skeletons had been measured in inches instead of centimeters and millimeters, then the correlation between stature and metacarpal length for these 10 skeletons would go up, go down, or stay the same. Pick one and explain.

b. One of these skeletons (identified by the X) had a metacarpal size of 52 mm and a height of 183 cm. If the height of this skeleton had been misrecorded as 153 cm, then the correlation between stature and metacarpal length for these 10 skeletons (including the misrecorded value) would go up, go down, or stay the same. Pick one and explain.

c. Using the data in the scatterplot above (i.e., without the error mentioned in part b), a researcher gets the following output for the regression of stature on metacarpal length:

```
Dependent variable is:     stature
No Selector
R squared = 78.5%    R squared (adjusted) = 75.8%
s = 3.983  with  10 - 2 = 8  degrees of freedom

Source          Sum of Squares    df    Mean Square    F-ratio
Regression      463.208           1     463.208        29.2
Residual        126.892           8     15.8615

Variable    Coefficient    s.e. of Coeff    t-ratio    prob
Constant    93.9906        14.62            6.43       0.0002
metacarpal  1.70736        0.3159           5.40       0.0006
```

A new metacarpal bone, which is 45 mm long, is found at an archeological dig. An investigator wants to use the data from the 10 skeletons mentioned above to make a prediction about the height of the person this new metacarpal bone came from. For the new metacarpal bone that was found, you would expect it to come from a skeleton that was _____ cm tall. Fill in the blank and explain.

118. The weights of 148 sets of twins born at the MetroHealth Medical Center in Cleveland, Ohio were recorded in 1986 and 1987. How strongly is the weight of the first born associated with the weight of the twin? A scatterplot is shown below (all weights are in kilograms).

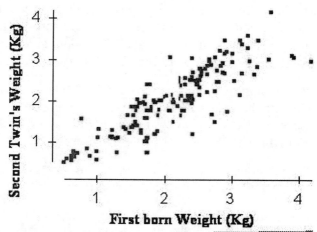

a. The correlation between the weight of the first born and the weight of the twin is about
   i) −0.87
   ii) −0.36
   iii) 0.36
   iv) 0.87
Explain your choice.

b. If the weights of the first born had been measured in pounds instead of kilograms, then:
   i. the value of the median weight of the first born twin would be _____.
   ii. the value of the standard deviation of the weights of the first born would be

   _____.

   iii. the value of the correlation between the two twins' weights would be

   _____.

   Fill in the blanks from the possible choices listed below (*Note*. There are about 2.2 pounds in one kilogram). You may use an answer more than once.
   (1) be multiplied by 2.2   (2) be divided by 2.2
   (3) stay the same   (4) be multiplied by 2.2 times the correlation

119. True or false: If there is a linear pattern to the data, then linear regression can be appropriately used for extrapolation.

120. True or false: If there is a non-linear relationship between x and y, then the correlation will always be zero.

121. True or false: The correlation r measures both the direction and strength of a straight-line relationship.

EESEE Exercises

122. **EESEE Story** *Nutrition and Breakfast Cereals.* This story details a study of the nutritional content of popular breakfast cereals. Read through the introduction to this story and open the data file called "Cereals." This data gives the nutritional information from the box labels of 77 brands of breakfast cereal.

   a. Examine the amount of sodium in the cereals. Make a histogram and describe its shape. Calculate the values needed for the five-number summary and make a boxplot. Does the five-number summary do a good job of describing the distribution in this case?

   b. Examine the relationship between the amount of potassium in the cereals and the amount of dietary fiber. What is the correlation?

   c. Suppose a new breakfast cereal comes on the market with 300 milligrams of potassium per serving. Would it be appropriate to use the regression method to predict the amount of dietary fiber in a serving of this cereal? If yes, what is the prediction? If no, explain why not.

123. **EESEE Story** *Hubble Recession Velocity.* In 1929 the astronomer Edwin Hubble investigated the relationship between the distance from Earth (in millions of light years) and the recession velocity (in kilometers/sec) of 24 galaxies. Read through this story's protocol. Hubble theorized that the relationship should be approximately linear. The data from Hubble's investigation are in the data file called "Hubble."

   a. Suppose you find a galaxy with a recession velocity of 500 km/sec and want to estimate that galaxy's distance from Earth using Hubble's data. Which variable would you choose to be the y variable, and which would you choose to be the x variable? Explain.

   b. Use the computer to fit the regression line for 'distance' on 'recession velocity'. How far away from Earth do you predict the galaxy from part a to be?

124. **EESEE Story** *Blood Alcohol Content.* How much does drinking beer increase the alcohol content of your blood? Read the introduction and protocol for this story. This question was addressed in an experiment at the Drackett Towers dormitory on The Ohio State University campus on February 20, 1986. Sixteen students volunteered to take part in the experiment. Before the experiment each of the subjects blew into a Breathalyzer to show that their blood alcohol content (BAC) was at the zero mark. The student volunteers then drank a varying number of 12 ounce beers (between one and nine). How much each student drank was assigned by drawing tickets from a bowl. About 30 minutes later, an officer from the OSU Police Department measured their BAC using the Breathalyzer machine. Data from this experiment is in the datafile called "bloodalc." Details of the variables are given in the results section of the story.

a. Suppose you want to estimate how a person's Blood Alcohol Content is affected by the number of beers they drink. Make a scatter plot of 'BAC' versus 'beers'. Which variable did you choose to be the y-variable and which did you choose to be the *x* variable? Explain.

b. Use the computer to find the correlation between 'BAC' and 'beers'. Is the correlation coefficient an appropriate measure of the strength of the association between BAC and beers? Explain briefly.

c. Use the computer to fit the regression line for 'BAC' on 'beers'. If a student drinks five beers, on average what do you predict the student's BAC will be? Show your work.

d. Would the regression method be as accurate for predicting BAC for a person who drinks 15 beers? Explain.

Online Problems for Chapters 14 - 15

125. To understand the idea of regression and correlation try the applet at the following Web site : www.whfreeman.com/applet/corr.html
Read through the directions to the applet. Notice that you can add points to the scatter plot just by clicking. The correlation of the points will appear in the upper-right corner. To clear the points you can double click the trash can icon.

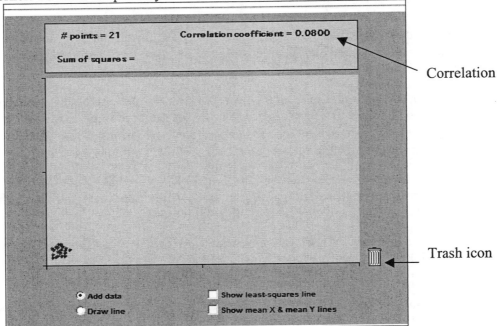

a. Create a scatter of points in the lower-left corner that has a correlation that is near zero.

b. Now add a single point to your scatterplot in the upper-right corner. Click and drag the point to different places on the scatterplot. How much can you change the correlation by manipulating this single point?

c. Clear your scatter plot using the trash icon. Create a new scatterplot that has a straight line of points and a correlation that is near 1.

d. Now add a single point to your scatterplot in the upper-right corner. Click and drag the point to different places on the scatter plot. How much can you change the correlation by manipulating this single point?

e. Use what you have learned above to answer the following question:
An outlier in a scatterplot can:
i) increase the correlation.
ii) decrease the correlation.
iii) either increase or decrease the correlation.
iv) Have no influence on the correlation

126. What is the best line that fits a pattern of points? The least squares line is the line that minimizes the squared vertical distance from the points and the line. How well can you determine this line? You can use the applet at www.whfreeman.com/applet/corr.html to experiment with determining this line.

a. Open the applet. Create a scatterplot that has a linear pattern and a correlation around 0.7.

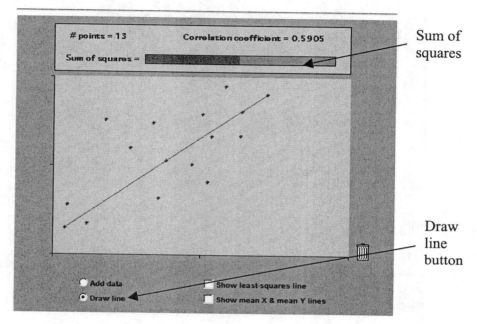

b. Click the "Draw Line" radio button. You can now click and drag in the scatter plot to create a line. The resulting sum of squares for your line will be indicated by the bar. The minimum possible sum of squares is indicated by the red part of the bar. Try moving the line you have created to try to reduce the green part of the bar.

c. When you have the line that you think is best you can click on "Show least-squares line" to see the actual best fit line. How did your line compare?

d. Draw a new set of points and see if you can draw a line close to the least squares line on your first attempt. Was the line you drew centered correctly? Was its slope too steep or too shallow?

127. How well can you match correlations to their scatterplots? To find out try the following online applet:
www.stat.uiuc.edu/courses/stat100//java/GCApplet/GCAppletFrame.html

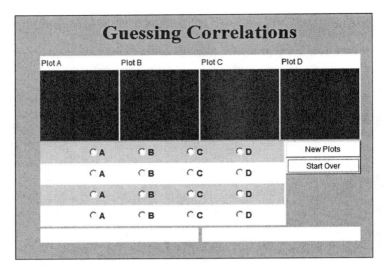

a. Click the "New Plots" button. The applet will present four scatterplots and four correlations.

b. Examine each plot and pick the correlation coefficient that matches the scatterplot. When you have made your guesses click the "Answers" button to find out if you are correct.

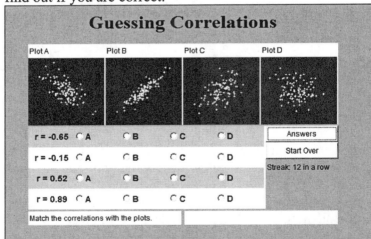

c. You may continue to generate new plots. Try to achieve a streak of at least 20 in a row.

114

Additional Problems for Chapters 17 - 20

128. True or false: In tossing a fair coin, the probability that you get Heads on between 49% and 51% of the tosses decreases as the number of tosses increases.

129. True or false: In tossing a fair coin, the probability that you get Heads on *exactly* half of the tosses increases as the number of tosses increases.

130. About 48% of all infants are girls. The maternity ward of a large hospital handles 1000 deliveries per year. A smaller hospital in the same city records about 50 deliveries per year. At which hospital is it more likely that between 46% and 50% of the babies born there this year will be girls? Explain briefly.

131. Two factories each produce television sets for the same company, and the chance of producing a set that requires service during the 90-day warranty period is 1% at both factories. The first factory makes 1573 televisions per week, whereas the second factory only produces 128.

    a. At which factory is it more likely that over 2% of this week's production will require service during the 90 days of the warranty? Explain.

    b. Does your answer to part a mean that it would be wise to buy a television made at the other factory? Explain.

132. On December 26, 1991, the Associated Press reported an unusual occurrence at the Canton-Postdam Hospital in Postdam, New York. It seems that between December 15 and December 21, all 16 newborns at the hospital were boys.

    a. Would the births of 16 consecutive boys at a different hospital in another small town in the United States be just as newsworthy? Do you think this Associated Press report is very unusual? Explain.

    b. It was also explained that on the two days before the report appeared, all five newborns at the Canton-Postdam Hospital had been girls. Does this show the Law of Averages in action? Explain.

133. Ten thousand lottery tickets, with various six-digit numbers printed on them, are placed in a steel drum and the winning ticket is drawn out at random. Suppose there is a 0.15 probability for the winning ticket to be the number 327849 or less. Also, the probability is 0.10 for the number to be 856183 or more. True or False: The probability that the number drawn will be between 327849 and 856183 (inclusive) is 0.75.

134. Assume that 20% of the cars in Wheaton, Illinois are green. An observer stands at the intersection of Butterfield Road and Park Boulevard in Wheaton for an hour and sees 898 cars traveling along Butterfield and 243 cars traveling along Park.

    a. In which direction is it more likely that more than 25% of the cars seen will be green?

        i) It is more likely among the 898 cars traveling on Butterfield Road.

        ii) It is more likely among the 243 cars traveling on Park Boulevard.

        iii) It is just as likely among either group of cars.

        Pick one and explain.

    b. Suppose that you watch 20 cars in a row traveling along Butterfield Road at this intersection and not a single one is green. What is the probability that the next car will be green?

    c. In answering parts a and b, what assumption have you made about the relationship between the colors of cars?

135. You roll a fair six-sided die and the die lands with a 3 face up. What is the probability that second roll also results in a 3? Pick one and explain your choice.

        i) 1/36 because there are 36 possible outcomes when you roll a die twice and getting two 3s is only one of them.

        ii) Slightly less than 1/6 because a second roll of the same thing does not happen often.

        iii) Slightly more than 1/6 because 3s are more likely with this die.

        iv) 1/6 because the two rolls do not influence each other.

136. A bag contains 10 poker chips of which four are blue, four are red, and two are white. We reach into the bag and pull three chips at random and find they are all red, and we do not put them back into the bag. We are going to pull another chip from the bag. It is most likely to be:

        i) red because they tend to be pulled from the bag more often.

        ii) blue because there are more of them in the bag.

        iii) red because the draws will be independent.

        iv) Since there are three colors, all three are equally likely.

    Explain your choice.

137. A county selects people for jury duty at random from a list of one million registered voters which happens be evenly divided in terms of gender (i.e., exactly 50% of the county's registered voters are women). Which of the following is true and which is false? Explain your choice.

        i) Out of the next 2000 names drawn, it is likely that exactly 1000 will be women.

        ii) Out of the next 2000 names drawn, it is likely that the percentage of women will be around 50% – but may be off by a percent or two.

138. Which of the following describe independent events? Explain your choice.
   i)   Whether you get "Heads" on the first toss and whether you get "Heads" on both of the first two tosses of a coin.
   ii) Whether you get "Heads" on the first toss and whether you get "Heads" on the second toss of a coin.

139. Forty percent of the customers who purchase gas at an EXXON station will also buy something besides gas. Which of the following is the most likely to happen?
   i)   A majority of the next 20 customers who purchase gas will also buy something else.
   ii) A majority of the next 100 customers who purchase gas will also buy something else.
   iii) Both i and ii are equally likely.
   Pick one and explain your reasoning.

140. Twenty percent of the subscribers to a daily newspaper receive the Sunday issues only; 35 percent receive the paper daily but not on Sunday. A subscription may be Sunday only, Monday–Saturday only, or every day. The editor of the paper picks a household at random from the list of subscribers.

   a. What is the probability that this household receives the paper every day (Sundays included)?

   b. What is the probability that this household receives the paper on Sunday?

141. We have poker chips with the numbers 1 through 40 painted on them. We place these chips in a bag and randomly select six of the chips. Which of the following is more likely:
   i)   Getting the numbers 1, 2, 3, 4, 5, and 6
   ii) Getting the numbers 2, 4, 6, 8, 10 and 12
   iii) Getting the numbers 10, 14, 20, 34, 35, and 38
   iv) All are equally likely.

142. A spinner can land in eight possible positions so that all eight have the same probability of coming up. Each position must have probability
   i)   between 0 and 1, but can't say more.
   ii) between -1 and 1, but can't say more.
   iii) 1/2.
   iv) 1/8.

143. We roll a fair six-sided die five times. Which of the following sequences is most likely to result from this experiment in this order:
    i)   2, 5, 1, 6, 6.
    ii) 6, 2, 5, 1, 6.
    iii) 3, 3, 3, 4, 4.
    iv) All of these sequences are equally likely.
    v)   Sequences i and ii are more likely than iii, but are equally likely.

144. Thirty percent of the customers at a hardware store pay using a credit card. Consider the following two probabilities:

    Probability I – When 100 customers are selected at random – the probability that between 20 and 40 of them will use a credit card.

    Probability II – When 10 customers are selected at random – the probability that between 2 and 4 of them will use a credit card.

    Which is true? Pick one and explain briefly.
    i)   The probability described in **I** is larger.
    ii) The probability described in **II** is larger.
    iii) The probabilities described in **I** and **II** are the same.

145. When a thumbtack is flipped, it may land point up (⊥̆) or point down (⌐). The probability that it lands point up is 0.69.

    a. What is the probability that it lands point down?

    b. Consider the following two probabilities:

        Probability I – When a thumbtack is flipped 1000 times independently; the probability that it lands point-up on more than 75% of the tosses.

        Probability II – When a thumbtack is flipped 20 times independently; the probability that it lands point-up on more than 75% of the tosses.

    c. Which is true? Pick one and explain your choice.
        i)   The probability described in **I** is larger.
        ii) The probability described in **II** is larger.
        iii) The probabilities described in **I** and **II** are the same.

    d. Suppose that the thumbtack is flipped independently 10 times and it lands point up on all 10 of the tosses. True or False: The probability that it will land point up on the next toss is 0.69.

146. In a city with no cars of mixed color, 30% of the cars on the road are brown. Forty percent are either brown or green, and 10% are white.

   a. What is the probability that a randomly selected car is green?

   b. What is the probability that a randomly selected car is not brown, green, or white?

147. A soda machine sells Coke, Diet Coke, Sprite, and Root Beer. Twenty percent of the customers who use this machine buy Diet Coke, 20% buy Sprite, and 10% buy Root Beer.

   a. What is the probability that the next customer to buy soda from this machine will buy a Coke?

   b. What is the probability that the next customer will not buy a Sprite?

148. A college student is taking a multiple choice test that has 10 questions. Each question has 4 possible answers. We would like to simulate what would happen if the student guessed randomly on each question. Specifically we would like to examine the number of correct answers. Which of the following methods is appropriate to simulate this situation?
   i)   Repeatedly flip a fair coin 10 times, keeping track of the number of heads.
   ii) Repeatedly use a random number generator to generate 10 random numbers between 1 and 4. Count the number of 4s.
   iii) Repeatedly roll a six-sided die and count the number of 4s.
   iv) Both ii and iii would be appropriate.

149. A gambler would like to test his latest strategy for the game roulette. In roulette a wheel is spun that has 38 numbered slots. Of these slots 18 are red, 18 are black, and 2 are green. The wheel is spun and a ball falls into one of these slots at random. His strategy is set up that he will win $2 if the wheel lands on red, and he will win $10 if the wheel lands on green. If the wheel lands on black he wins $0.
   a. Write a probability model for the amount that the gambler would win.

   b. The gambler proposes to simulate the result by generating a random number between 0 and 9. If the number is 0 he will add $10 to his total. If the number is between 1 and 4 he will add $2 to his total. If the number is between 5 and 9 he will add $0 to his total. He will generate 1000 random numbers. Is this a legitimate way to simulate his system? Explain.

150. The registrar at a large Midwestern university knows that of undergraduate students at the university, 30% are freshmen, 30% are sophomores, 20% are juniors, and 20% are seniors. The registrar is going to choose a focus group of 10 undergraduates by random selection. However, the registrar is worried that the group might end up being all upperclassmen (Juniors and Seniors) and would like to simulate the random selection to see how often this might happen. The registrar's staff puts together several different proposals as to how this simulation could be carried out. Consider each of these and determine which would be a valid simulation of this situation and which would not.

a. For each run of the simulation, roll a four-sided die (see problem 18.7 in the text for a description) 10 times. If the die lands on 3 or 4, count them as upper classmen. Record the total number of upperclassmen in the 10 rolls. Repeat the run 1000 times to determine how often we would have a focus group of all upperclassmen.

b. For each run of the simulation roll a 10-sided die 10 times. If the die lands on a 1, 2, 3, or 4 count that as an upperclassman. Record the total number of upperclassmen in the 10 rolls. Repeat the run 1000 times and observe how often we would have a focus group of all upperclassmen.

c. For each run of the simulation use a computer to generate a random number between 1 and 10. Use the random number as the number of upperclassmen in the focus group. Repeat the run 1000 times and observe how often we would have a focus group of all upperclassmen.

Online Problems for Chapter 17 - 20

151. The idea of probability is often difficult to grasp for some students. An applet at www.whfreeman.com/applet/probability.html can give an indication of how probability works. The applet simulates flipping a coin and keeps track of the proportion of heads and tails. Open this Web site and try the exercises below.

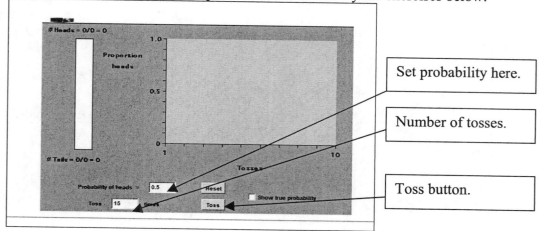

a. Will you always get five heads out of 10 flips of a fair coin? We can simulate this by setting the number of tosses to 10 and clicking toss. What was the resulting proportion of heads? Click reset and repeat your simulation. What was the result? Reset and repeat your simulation a total of 10 times, recording the proportion of heads. What was the highest percentage of heads you saw? What was the lowest percentage of heads you saw?

b. How would our results differ if we had flipped the coin 40 times on each simulation? Change your number of tosses to 40 and click toss. Reset and repeat your simulation a total of 10 times, again recording the proportion of heads. What was the highest percentage of heads you saw? What was the lowest percentage of heads you saw?

c. Compare your results for parts a and b. Were the proportions more varied in part a or in part b? Summarize your results by completing the following sentence: If a coin flip is repeated many times the proportion of heads will approach _____%. In fact, if we flip a coin ____ times we will find that the proportion of heads tends to be less varied than if we flip a coin ____ times.

Additional Problems for Chapter 21

152. A large university records the number of credit hours taken during the semester by its students. At that university students taking 13 credit hours or more are considered full time, but there are some part time students that may be taking as few as 1 credit hour per semester. Students are not allowed to take more than 20 credit hours, but most students take at least 14 hours. For the most recent semester the university produced the following summary for the number of credit hours.

Mean	Median	Min	Max	Q1	Q3	St Dev.
15.5	16	1	20	14	17	2.1

a. We are going to randomly select 90 students from this university. What is the probability that the average number of credit hours taken by these students would be over 15 hours? Show your calculations.

b. Would this data be considered to be skewed left, skewed right, or symmetric? Explain how you know. Does this make your answer to part a invalid? Explain why or why not.

153. A large state university in the midwest records many factors about their students. One of the items they record is the distance from the university to the student's permanent address. For undergraduates the vast majority (91%) of the students at this university have a permanent address within 100 miles of the university although there are some students from other parts of the United States and just a few from other countries, making some of the distances very large (maximum 7023 miles). The average distance for undergraduate students is 85 miles and a standard deviation of 180 miles.

a. For undergraduates at that university, would the distribution of the distance to students' permanent addresses be skewed left, skewed right, or symmetric? Explain how you know.

b. We randomly select 100 students from this population. What is the probability that the average distance to their permanent addresses is less than 50 miles? Show your work and explain why your calculations are valid.

154. A large company has found that 40% of its employees own their homes.

    a.   If we randomly select a sample of 20 employees, what is the probability that the sample proportion that own their homes is over 50%?

    b. If we randomly select a sample of 40 employees, what is the probability that the sample proportion that own their homes is over 50%?

    c. If we randomly select a sample of 20 employees, what is the probability that the sample proportion that own their homes is between 30% and 50%?

155. True or false: Since the nature of the population does not matter, the sampling distribution of a sample proportion will approximately follow the normal distribution, even for a convenience sample.

156. A vegetable supplier makes deliveries of iceberg lettuce to a chain of grocery stores. The heads of lettuce weigh an average of 16 ounces with a standard deviation of 4 ounces. If 100 heads of lettuce are picked randomly from this group, what is the chance that they will have an average weight of less than 15.8 ounces?

157. A local utility company has found that the usage of water by single family residential homes in the city averages 12,700 gallons per month with a standard deviation of 8100 gallons. We randomly select 50 homes from this city.

    a. Explain why this mean and standard deviation make it unlikely that the distribution of water usage is normally distributed.

    b. What is the chance that the average water usage for these 50 homes is over 15,000 gallons?

    c. What is the chance that the average water usage for these 50 homes is less than 11,000 gallons?

    d. What is the chance that the average water usage for these 50 homes is between 13,000 and 17,000 gallons?

158. In a simple random sample of 225 UCLA undergraduates, 45 say they plan to go on to graduate school. Find an 80% confidence interval for the proportion of all UCLA undergraduates who plan to go on to graduate school.

159. Forty-nine compact car engines of the same type are randomly selected using an SRS and independently tested by the EPA for nitrogen oxide emissions. The test results show an average of 1.5 grams per mile with a standard deviation of 0.84 grams per mile. Find a 90% confidence interval for the average grams of nitrogen oxide emitted per mile for this type of engine.

160. An experiment was reported in the *American Journal of Public Health* regarding a needle exchange program in San Francisco. In an effort to slow the spread of the HIV, the virus that causes AIDS, among intravenous drug users, this program exchanges clean needles for used needles on a one-to-one basis (an individual patron usually exchanges about 10 needles at a time). The program directors would like to know what percentage of the clean needles they give out eventually come back for exchange. To estimate this percentage they handed out 4239 color coded needles of which 2593 (61.2%) were returned within nine weeks. Using this data, the following 80% confidence interval is created for the percentage of needles returned for

exchange: $0.612 \pm 1.28 \left(\sqrt{\dfrac{(0.612)(0.388)}{4239}} \right)$. Is this confidence interval appropriate?

Explain why or why not.

161. An investigator wishes to estimate the percentage of Sacramento, California voters who favor an increase in the sales tax to fund education. A simple random sample of 275 voters is taken and it is found that 47% of the sampled voters are in favor of the proposal.
   a. The interval 47% ± 2% is approximately a(n)_____ confidence interval.
       i)   95%
       ii) 80%
       iii) 68%
       iv) 50%

Pick one and explain.

   b. If, instead of the interval above, we used the interval 47% ± 4%, then the level of confidence would
       i)   go up.
       ii) go down.
       iii) stay the same.
       iv) We can't tell based on information given.

162. One hundred and fifty students attend the lecture of a statistics course on the first day of class. The course instructor wants to estimate the average height of the students in the room and asks the nine students in the front row to write down their heights. The average height of these nine students turned out to be 64 inches with a standard deviation of 3 inches.

   a. What is the parameter of interest? What is the sample statistic?

   b. Would it be appropriate to use the normal distribution and the formula $\bar{x} \pm z^* \dfrac{s}{\sqrt{n}}$

       to make a confidence statement about the average height of the students in the class based on the data given in the problem? If yes, construct the appropriate confidence interval. If no, explain what's wrong.

163. Indicate whether the following statements are true or false?

   a. When the data is collected using a large SRS, then the sampling distribution of a sample proportion will approximately follow the normal distribution.

   b. Because $\frac{95}{68} \approx 1.4$, 95% confidence intervals are usually about 1.4 times as wide as 68% confidence intervals.

   c. A 95% confidence interval for a population proportion, p, will be shorter when the sample size is 1000 than when the sample size is 500.

164. In a certain city, there are 1 million homes. As part of an environmental status survey, it was desired to estimate the proportion of homes in this city that contain lead-based paints. A simple random sample of 1600 households revealed that 160 homes had lead-based paints in at least one room. In this problem,

   a. What is the population?

   b. What is the sample?

   c. What is the parameter of interest?

   d. What is the sample statistic?

   e. Find an 80% confidence interval for the proportion of the city's homes that contain lead-based paint.

   f. Which of the following would *decrease* the margin of error in the interval you constructed in part e; which would *increase* the margin of error; and which would have little effect on the margin of error? Explain.
      i.   If you had a random sample of 400 households instead of 1600.
      ii. If there were two million homes in the city instead of one million.
      iii. If you created a 68% confidence interval instead of the 80% confidence interval.

165. A random sample of 1000 people who signed a card saying they intended to quit smoking on November 20, 1995 (the day of the "Great American Smoke-out") were contacted in June 1996. It turned out that 210 (21%) of the sampled individuals had not smoked over the previous six months.

   a. Specify the population of interest, the parameter of interest, the sample, and the sample statistic in this problem.

   b. Make a 90% confidence interval for the percentage of all people who had stopped smoking for at least six months after signing the non-smoking pledge.

166. The city planning department wants to know what percentage of homes in the city have trees taller than 10 feet on the surrounding property. A random sample of 400 homes is taken and visits to these homes find that 300 of them (75%) have trees taller than 10 feet. In this situation,

    a. What is the population?

    b. What is the sample?

    c. What is the parameter of interest?

    d. What is the sample statistic?

    e. Make a 90% confidence interval for the percentage of all homes in the city that have trees taller than 10 feet.

167. A study in the *Journal of Zoology* examined the migration and survival of house mice on a small island off the East coast of Scotland. Traps were set up in October of 1971 to obtain a sample of 247 mice, which were tagged and released. In April of 1972 a saturation grid of traps was laid out for a period of five days to try to capture *all* of the mice in the area. It was found that 84 (34%) of the previously tagged mice were captured in this group. The investigators presented 34% as an estimate of the chance that house mice survived the winter of 1971 – 72.

    a. What assumptions about house mice are being made by the investigator in order for this to be a reasonable estimate? Do you think the assumptions are likely to be true?

    b. If the assumptions were true, what would be a 90% confidence interval for the percentage of mice on the island that survived the winter? Can you think of a better description of the parameter that is being estimated?

    c. The experiment was repeated the next year with 141 mice tagged in October and 75 (53%) of these recaptured the next April. Make a 90% confidence interval for the percentage of mice who survived the winter of 1972 – 73.

    d. Do the two intervals from parts b and c overlap? Comment.

168. Two newspapers report the results of the *same* Gallup Poll regarding the percentage of people who favor term limits for Senators. The first newspaper uses the poll to present a 95% confidence interval for this percentage, whereas the second newspaper presents an 80% confidence interval. True or False: The confidence interval presented by the first newspaper will be wider than the interval reported by the second paper.

169. One thousand students are randomly selected from the list of those currently registered at Harvard University, and they each report how many miles they rode in an MTA bus during the past week. Because many students did not ride a bus at all, a histogram of the values does not look like the normal curve. True or False: Even though the histogram did not follow the normal curve, it is still possible to use the normal curve to make a confidence interval for the average number of miles that all Harvard students rode on MTA buses last week.

170. The Transportation Department of a large city wishes to estimate the percentage of all street lamps in the city with burnt out bulbs. A random sample of 400 street lamps are examined, and 10% of them are found to have burnt-out bulbs. A 68% confidence interval for the percentage of the city's street lamps with burnt out bulbs is given by:
    i)   10% ± 1.5%
    ii)   10% ± 3%
    iii) 10% ± 10%
    iv)   10% ± 30%
       Explain your choice.

171. A nationwide chain of electronics stores wishes to know the average income of the households in Franklin County, Ohio before they decide to build a new store there. A random sample of 100 households is taken and the income of these sampled households turns out to average $35,000 with an SD of $20,000. Which of the following is a 68% confidence interval for the parameter of interest?
    i) $35,000 ± $200,000
    ii) $35,000 ± $20,000
    iii) $35,000 ± $2,000
    iv) $35,000 ± $200

172. A newspaper wants to determine whether its readers believe that government expenditures should be reduced by cutting welfare programs. They provide a telephone number for readers to call and give their opinions. Based on 1434 calls that they received, they report that 1190 (82.9%) of their readers believe that welfare payments should be reduced.

    a. Identify the population of interest in this situation.

    b. What is the parameter of interest?

    c. Would it be appropriate to calculate the margin of error for estimating the true proportion who believe that welfare should be cut? If so, calculate a 99% confidence interval. If not, explain why not.

173. The Gallup Poll conducted a survey to determine how American adults feel about their personal finances. The poll, conducted January 17 – 23, 2005, found 52% of the people surveyed believed they would reduce their debt in the next six months. On the other hand, only 33% of these same people had been able to decrease their debt in the past six months. According to the report at Gallop.com: "For results based on the total sample of 2007 national adults, one can say with 95% confidence that the maximum margin of sampling error is ±2.4 percentage points."

    a. If Gallup had reported the margin of error for the sample percentages using an 80% confidence level, it would be:

      i) greater than 2.4%.

      ii) less than 2.4%.

      iii) still 2.4% as reported above.

      Explain your choice.

    b. If Gallup took a second poll of another group of 2007 people that same week and asked the same questions, which of the following is most likely to change by two percentage points when the results are reported for this second poll? Pick one and explain.

      i) the sample statistic

      ii) the parameter

      iii) the margin of error

174. The director of student affairs at a large university was interested in the health of students at the university. He conducted a survey in which 200 randomly selected students from this university were asked if they had ever had a sexually transmitted disease (STD). The survey found that 38 of the students answered yes to the question.

    a. What is the parameter of interest in this study?

    b. The director has asked that you calculate the margin of error in this situation using 99% confidence. Calculate this margin of error.

    c. After the director reported the results of his study, a critic of the study said, "This result clearly would underestimate the proportion of students who had had STD's. Many students will not honestly answer a sensitive question such as this." Would this criticism be a sampling or a non-sampling error? Explain why.

    d. The director responded to this criticism with the following: "Issues such as students not answering honestly are not a problem because we have calculated the margin of error for this study at 99% confidence instead of the usual 95%." Does this adjustment of the confidence level adjust for the problem mentioned? Explain why or why not.

175. A news organization conducted a study to estimate the average amount of time (in hours) local residents spend watching news channels like CNN, Fox News, and MSNBC on television each week. This study interviewed 200 randomly chosen local residents and found the following summary statistics:

Mean	Median	Q1	Q3	Min	Max	St. Deviation
1.64	0.75	0.25	2.3	0	21.2	2.42

a. This distribution is skewed the right. Explain how we know this from the information given.

b. Since this distribution is skewed to the right, would it be appropriate to calculate a 95% confidence interval for the mean amount of time residents in this area watch news channels? If so, calculate the interval. If not, explain why not.

176. Fifteen percent of all food products imported into the Netherlands are in violation of the maximum pesticide residue limits established for food products in that country. The Inspectorate of Health Protection of the Food Inspection Service in Amsterdam takes random samples of food products being imported into the Netherlands and inspects them for pesticide residues. Next year, the Inspectorate plans to analyze a random sample of about 2500 items.

a. What is the probability that at least 14% of these samples will be in violation of the maximum pesticide residue limits? Show your work.

b. Which of the following is the most likely to happen:
    i)   At least 20% of the next 20 items inspected violate the standards.
    ii) At least 20% of the next 200 items inspected violate the standards.
    iii) Both i) and ii) are equally likely
         Pick one and explain your reasoning.

177. A market research company is interested in whether or not households in Franklin County Ohio have cable television. Using the county auditors list of households, the researcher randomly selects 100 homes and contacts the residents. Of the 100 homes it is found that 62 have cable television.

a. What is the parameter of interest in this case?

b. Would it be appropriate to calculate a confidence interval for the parameter? If so, calculate an 80% confidence interval. If not, explain why not.

178. A large university was interested in the proportion of students who purchased their textbooks online rather than at the bookstore. A random sample of 200 students found that 132 purchased their books online. Describe the parameter of interest and find a 90% confidence interval.

EESEE Exercises for Chapter 21

The following exercises make use of stories in the Electronic Encyclopedia of Statistics Examples and Exercises or EESEE (pronounced ee-zee). EESEE is included in the E-STAT Pack that accompanies this workbook. You can also access EESEE at www.whfreeman.com/eesee.

179. **EESEE Story** *Columbus' 1993 Election Poll.* In 1993 the city of Columbus Ohio was building up to a local election. Prior to the election, the local news paper *The Columbus Dispatch* conducted a poll by sending out a mock ballot to randomly selected individuals. Open this story and read through the Protocol and examine the Results.

    a. Use the Poll results to calculate a 95% confidence interval for the proportion of votes for C. Lazarus in the city council election.

    b. Compare the results of your interval with the actual election results. Does your interval contain the true proportion of the voters who voted for Lazarus in this election? Explain.

Online Problems for Chapter 21

180. The idea of a sampling distribution is that different samples will produce different statistics. If we take many samples we will see many different statistics. If we take those different statistics and put them into a histogram, we can visualize the sampling distribution of the statistic. It is the sampling distribution that allows us to determine many of the elements that we see in research, such as margins of error. To help visualize the sampling distribution we can make use of an online applet. To see this, open the Web site www.ruf.rice.edu/~lane/stat_sim/sampling_dist

Allow the page to fully load (this may take a moment). When the page has loaded, click the button marked "Begin".

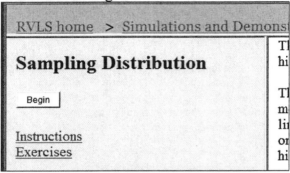

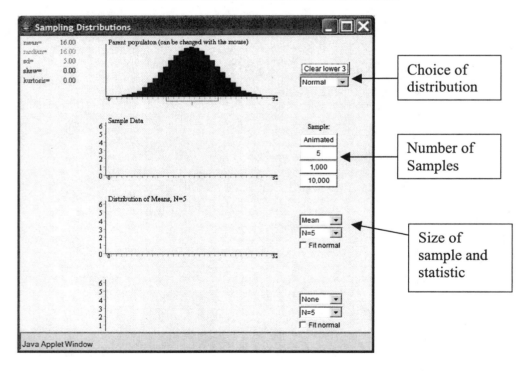

a. Assume our population is normally distributed. We would like to take a sample of five items from this population. To do this set the choice of distribution to "Normal." Next set the statistic to "Mean" and the sample size to N = 5. Click "Animated" to see each individual value in the first sample appear on the section marked "Sample Data." A blue mark will appear on the panel marked "Distribution of the Mean." What was the mean of this sample? Did you have any values in the sample that were near zero or near 32?

b. Take a second sample by again clicking on "Animated." What was the mean of this second sample? Did it contain any values near zero or 32? Continue to take samples by repeatedly clicking "Animated" until you have taken 10 samples.

c. Rather than repeatedly clicking the "Animated" button we can click sample numbers of 5, 1000, or 10,000. Use the 1000 button to quickly take 1000 additional samples. Observe the shape of the resulting distribution. How does this shape compare with the shape of the original Parent Population? How does the variability of the distribution compare with the original Parent Population?

d. **Larger sample sizes:** Click "Clear lower 3" to clear out the values from your previous experiment.
   i. Again set the statistic to the mean and the sample size to N = 5. In the second panel set the statistic to the mean and the sample size to N = 10. Sample 10000 times and observe the resulting distributions. How do the distributions compare in variability to each other and to the Parent Population?
   ii. Summarize your results by completing the following sentences: *A sample mean from a sample of N = _____ is less varied than the Parent Population but more varied than for a sample mean from a sample of N = ___.*

e. **Skewed populations:** If a population is skewed, what will the distribution of the sample mean look like? Will it be skewed or will it be symmetric? Change the shape of the distribution to "skewed."
   i. Take a single sample (N = 5) from this distribution. How many values in this sample were near the left half of the distribution? Where was the mean of the sample?
   ii. Set the first panel to N = 5 and the second panel to N = 25. Take 10,000 samples and describe the resulting distribution of the sampling distributions (symmetric, skewed right, skewed left, etc.). Which is more symmetric?
   iii. Summarize your results by completing the following sentences: *Even if a Parent Population's distribution is _____, the distribution of the sample mean is ____, and becomes more so as the sample size _____.*

133

181. The meaning of confidence is difficult for many people to grasp. An online applet located at www.whfreeman.com/applet/CI.html can give an animated illustration of this concept.

a. Open this applet. Select a confidence level of 80%. Click the "Sample" button. Observe that a line appears on the applet. The center mark on this line represents the mean of a random sample from the population. The line extends to the ends of a confidence interval. The green vertical line in the center of the graph indicates the location of the population mean μ. Does the confidence interval you created cover the population mean?

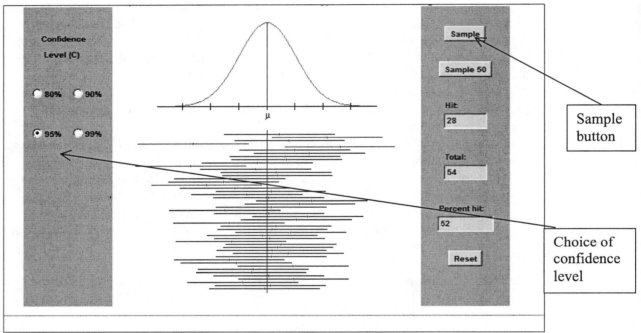

b. Repeat the sampling process until you have sampled 50 intervals. How many of these intervals covered the true population mean? What percentage of the intervals contain the true mean?

c. Change the confidence level to 90%. Did the lines representing your confidence intervals increase or decrease in width? Are there any intervals that contain the population mean now that did not contain it before? What percentage of the intervals contain the true mean?

d. Increase the confidence level again, this time to 95%. What percentage of the intervals contain the true mean?

e. Summarize what you saw in this applet by completing the following sentence: *When the confidence level increases, the width of the confidence interval* _____. *The confidence level indicates the percentage of possible confidence intervals that* _____ *the true mean.*

Additional Problems for Chapters 22 - 23

182. A gambler claims that when an ancient coin is tossed, tails come up too often. As an experiment, the coin is tossed 900 times. It comes up tails on 485 of these tosses. Is this strong evidence for the gambler's claim? Translate this into a statistical hypothesis-testing problem and carry out the test. Be sure to give the null and alternative hypotheses, the value of the test statistic, the p-value, and your conclusions.

183. Most people can roll their tongues, but many people can't. Whether or not a person can roll his tongue is genetically determined. Suppose we are interested in determining what fraction of students can roll their tongues. A recent news article claimed that 75% of people can roll their tongues, whereas others claim that the percentage is more than 75%. We get a simple random sample of 400 students and find that 317 can roll their tongues. Is this significant evidence that more than 75% of students can roll their tongues? Carry out the appropriate hypothesis test. Be sure to write down the null and alternative hypotheses, find the test statistic and the p-value, and state your conclusions.

184. The *Wall Street Journal* reported that 47% of soda drinkers drink a product produced by the Coca-Cola Company. A student government president at a southern university feels that students at his university drink a greater proportion of Coca-Cola products. A random sample of 216 students who drink soda were asked about the last soda they consumed. It turned out that 115 drank a Coca-Cola product. Does this indicate that the president's claim is correct, or is some other conclusion appropriate? Formulate this question as a statistical hypothesis test. Be sure to give the null and alternative hypothesis, the value of the test statistic, and the p-value. State your conclusion if we use a 5% level of significance.

185. A researcher is interested in whether birth order has an effect on the size of a child's vocabulary. A vocabulary test is given to 100 children, all age five and all the oldest in their families. Later, the same vocabulary test is administered to the 100 next oldest sibling of these children when they reach five years of age (so both children in each family take the same test at age five). In this study, the oldest child scored higher on the vocabulary test for 63 of the 100 pairs of children (while the younger child scored higher for 37 of the pairs). Is this strong evidence that order of birth has an effect on vocabulary size, or can these data be explained by chance? Carry out the appropriate hypothesis test. Be sure to write down the null and alternative hypotheses, find the test statistic and the p-value, and state your conclusions.

186. A test of hypothesis was conducted that resulted in a p-value of 0.02. This test is
    i)   significant at the 1% level.
    ii)   significant at the 5% level but not the 1% level.
    iii) not significant at the 5% level.
    iv)   We cannot tell without knowing the null and alternative hypothesis.

187. The EPA sticker for a particular model of automobile claims the car has average highway mileage of 35 miles per gallon. A consumer advocacy group takes a random sample of 30 of these cars and finds that they have an average mileage of 33.6 miles per gallon with a standard deviation of 3 miles per gallon. Do the results of this test provide sufficient evidence to conclude that the actual mileage of this model is less than 35 miles per gallon? Translate this into a statistical hypothesis and carry out the test of hypothesis. Be sure to specify the null and alternative hypotheses, the value of the test statistic, the p-value, and your conclusion.

188. A fast-food chain wishes to investigate the after cooking weight of their "Quarter-Pound" hamburger. The variation in the size of the hamburger patties is known to give a standard deviation of 0.015 pounds. However the machine that presses the patties may be improperly calibrated and produce patties that are systematically too large or too small. The company hires an independent laboratory to purchase 100 hamburgers at random times and independently weigh them using a procedure that is practically free of bias. The average of these weights is 0.245. Could the true average after-cooking weight of *all* of the "Quarter-Pound" hamburgers produced by this pressing machine still be 0.25 pounds, or is this strong evidence that the pressing machine is improperly set and the true average is not 0.25 pounds? Translate this into a statistical hypothesis-testing problem and carry out the test. Be sure to give the null and alternative hypotheses, the value of the test statistic, the p-value, and your conclusions.

189. The Capilano Fish Hatchery in British Columbia raises Coho Salmon and releases about one million per year as smolts into the Capilano River. The Hatchery would consider its program successful if it could prove that more than 5% of these fish survive to adulthood and return to the river to spawn. In 1983 an experiment was conducted in which 1000 smolts were tagged with a special coded wire. It turned out that 68 (6.8%) of the tagged fish returned as adults to the Capilano River. Is this strong evidence for the success of the program? Translate this problem into a statistical hypothesis-testing problem and carry out the test. Be sure to give the null and alternative hypotheses, the value of the test statistic, the p-value, and your conclusions.

190. A genetic model predicts that a cross between a certain red flowering plant and a white flowering plant is expected to yield offspring with a 0.5 probability of being pink flowering, independently of one another. One hundred seeds of such a cross are made, and only 45 turn out to give pink flowers. Is this strong evidence that the model is wrong and pink flowering plants arise less than 50% of the time? Carry out the appropriate hypothesis test. Be sure to write down the null and alternative hypotheses, find the test statistic and the p-value, and state your conclusions.

191. The Food and Drug Administration will allow a food product to be labeled "sodium free" only if there is strong evidence that it contains, on average, less than 5 milligrams of sodium per serving. A soft drink wishing to use a "salt free" label has its product tested. A laboratory takes 25 independent sample servings of the soft drink and measures them for sodium content using a technique that is practically free of bias. These measurements average 4.9 milligrams of sodium with a standard deviation of 1 milligram. Is this strong evidence that the new soft drink deserves the "salt free" label? Translate this problem into a statistical hypothesis-testing problem and carry out the test. Be sure to give the null and alternative hypotheses, the value of the test statistics, the p-value, and your conclusions.

192. How do flour beetles get into human pantries to infest dry goods like flour or oatmeal? Do they walk randomly until they find something, or are they guided by a sense of smell? In order to investigate this issue, a researcher constructs a "beetle track" like the one pictured below. The track consists of a circular band, divided into two sections. Oatmeal is placed in one section, and the other section is left empty. Individual beetles are released at the point, X, and it is observed whether they first touch the circular band in the oatmeal section or in the empty section. Altogether 100 beetles are independently run through this track, and it is observed that 66 of them reached the oatmeal section first. Is this strong evidence that the beetles are not just walking randomly? Translate this into a statistical hypothesis-testing problem and carry out the test. Be sure to give the null and alternative hypotheses, the value of the test statistic, the p-value, and your conclusions.

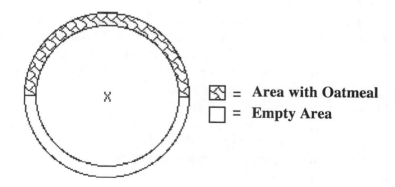

⬗ = **Area with Oatmeal**

☐ = **Empty Area**

193. Farms deliver dried fruit to a packing house in large wooden bins supplied by the packing house. When the fruit arrives it is weighed in the bins, then 140 pounds is subtracted, and the farmers are paid for each pound remaining. The 140 pounds is subtracted because that is how much the manufacturer of the bins claim they weigh on average. Recently, farmers have complained that these wooden bins really average less than 140 pounds and that they are not being paid the full value of their fruit. In order to examine this claim, a random sample of 64 bins are taken out of the warehouse and weighed. These 64 averaged 138 pounds with a standard deviation of 4 pounds.
   a. Is this strong evidence that the average weight of the bins in the warehouse is less than 140 pounds? Carry out the appropriate hypothesis test. Be sure to write down the null and alternative hypotheses, find the test statistic and the P-value, and state your conclusions.

   b. Make a 95% confidence interval for the average weight of the bins in the warehouse. Does your interval contain 140 pounds? Explain.

194. Microwave ovens emit electromagnetic waves similar to radio waves. A Federal Communication Commission regulation states that the average wavelengths of microwaves from ovens is supposed to be 4.81 inches. A group of engineers at a company that produces microwave ovens worry that their new design will result in ovens that emit wavelengths that are too long. To test an oven each engineer independently measures the wavelengths emitted and the average of their measurements is taken as the value for that oven. Sixteen of the ovens are tested in this way. The average of these 16 values is $\bar{x} = 4.9$ inches with a standard deviation of 0.1 inches.
   a. What does this evidence say about the engineers' theory? Translate this into a statistical hypothesis-testing problem and carry out the test. Be sure to give the null and alternative hypotheses, the value of the test statistic, the p-value, and your conclusions.

   b. Would the distribution of the measurements approximately follow the normal curve? Explain why or why not.

195. A metallurgist has invented a new tungsten alloy for use in the filaments of light bulbs that is supposed to burn brighter with the use of less energy. A company investigates the potential costs of producing the new type of bulb and decides that production would be economically feasible if the bulbs would last an average of longer than 2000 hours in continuous usage. Twenty-five of the new bulbs are tested and found to last an average of 2240 hours with a standard deviation of 600 hours. Is this strong evidence that the average life of all such bulbs is higher than 2000 hours? Translate this into a statistical hypothesis-testing problem and carry out the test. Be sure to give the null and alternative hypotheses, the value of the test statistic, the p-value, and your conclusions.

196. Does a student just guess on a multiple-choice test or do they actually understand the material? An instructor gives an exam that consists of 100 multiple-choice questions. Each question has four possible answers. The student in question completes all 100 questions. The result is that the student has 31 of the answers correct. Does this provide sufficient evidence to rule out random guessing by the student? Conduct an appropriate test of hypothesis, providing assumptions, hypotheses, test statistic, p-value, and conclusion.

197. A researcher is going to collect a random sample of 50 subjects and will calculate the sample mean. The population from which the sample will be taken has a standard deviation of 9. He will conduct a test of the null hypothesis that $\mu = 10$ against the alternative that it is greater than 10. He decides he will reject the null hypothesis if the sample mean ($\overline{x}$) is greater than 11. What significance level _ is he using? Explain how you obtain your value.

198. An ecologist for the state of Missouri has monitored the mercury level in the ground water near a large smelting plant. The ecologist took a random sample of 15 area wells and determined the mercury level in these wells. The measurements in mg/m^3 were: 0.99, 1.03, 1.07, 1.07, 1.20, 1.25, 1.35, 1.44, 1.60, 1.61, 1.77, 1.80, 1.87, 2.04, and 2.19. The average of these values is 1.49 and the standard deviation is 0.389. A court settlement requires that the smelter stop operations if the mercury level in the nearby ground water has an average of more than 1.25. The ecologist believes that measurements of mercury concentrations are approximately normally distributed. Conduct a test of hypothesis to see if the average is greater than 1.25mg/m^3.

199. The label on packages of a certain brand of granola indicates a net weight of 16 oz. A consumer group is suspicious that the boxes may typically contain less cereal than indicated. In order to check this theory, they weigh the contents of 64 randomly selected packages and find an average net weight of 15.7 oz. with a standard deviation of 1 oz. Is this strong evidence that the net weight on the labels is higher than the average for the contents of all packages? Carry out the appropriate hypothesis test. Be sure to write down the null and alternative hypotheses, find the test statistic and the p-value, and state your conclusions.

200. Indicate whether the following statements are true or false.

   a. A p-value is the probability that the null hypothesis is correct.

   b. If the p-value is small, then the sample statistic is a poor estimate of the population parameter.

   c. A p-value of 0.999 says that the null hypothesis is a reasonable explanation of the data.

   d. The p-value is calculated assuming the null hypothesis is true.

   e. When the p-value is 0.5, there is a 50-50 chance that the alternative hypothesis is true.

   f. A test statistic is used to measure the difference between the data and what is expected under the null hypothesis.

201. *Washington Post* columnist Sally Jenkins described an anecdote about golfer Tiger Woods' ability to detect subtle differences in golf equipment. Tiger Woods was sent six golf clubs to test. The six clubs looked identical, but one was heavier than the rest by two grams (about the weight of a dollar bill). Tiger Woods swung each club and quickly declared, "This one's heavier." He was right.

   a. If Tiger was really just picking a club at random, what would be the chance that he would guess the one that is heaviest?

   b. Suppose this basic experiment was repeated with 120 people who really could not tell the clubs apart (i.e., each person is given six identical looking clubs and just randomly guesses which one is heaviest). What is the chance that at least 25% of the people would guess correctly? Show your work.

   c. The experiment in part b is carried out for a random sample of 120 professional golfers, and 25% of them guess the correct club. A statistician carries out a significance test of the null hypothesis that these golfers were just guessing versus the alternative hypothesis that a greater proportion of professional golfers than expected under random chance can recognize the heaviest of six clubs.
   i) What are the null and alternative hypotheses in this situation?

   ii) What is the p-value for this significance test? Try to answer without doing any new calculations.

202. A researcher believes that the average height of trees in a particular section of forest is over 60 feet. To test the null hypothesis of $\mu = 60$ versus the alternative of $\mu > 60$ the researcher takes a random sample of 200 trees from this forest and finds a test statistic of $Z = -1.3$. The p-value for this test would be
   i) 9.68%   ii) 90.32%   iii) 19.36%   iv) It is impossible to tell.

203. **EESEE Story** *Kicking the Helium Football.* Many people have thought about ways to cheat at various sports. After a kicker achieved a very long punt in football, fans of the opposing team claimed the ball might be filled with helium. But can a helium-filled football really be kicked further? This EESEE story details studies that examine this question. Open the story and read through the introduction and the protocol of these studies.

  a. Most people believe that the helium-filled football would be kicked further than the air filled football. Based on this idea, what null and alternative hypothesis should we use to conduct a statistical test?

  b. Use statistical software to open the data set that contains the results of this experiment. Use the software to calculate the difference in each of the 39 pairs (air and helium) of kicks. Calculate the mean and standard deviation of these differences.

  c. Use the difference you found in part b to conduct a test of hypothesis to determine if the helium football was kicked significantly further than the air-filled football?

204. **EESEE Story** *Anecdotes of Significance Testing.* This story gives many quotes about p-values and significance testing. Read through the quote by Gaugh. Explain why his statement is incorrect.

205. **EESEE Story** *Therapeutic Touch.* Does the body emit a psychic energy? Practitioners of Therapeutic Touch believe that it does and that they can detect this energy field. Read through the protocol of this story.

  a. In this study the practitioners were given two choices and if they really can detect a psychic energy we would see them getting more than half of these selections correct. Based on this idea, what null and alternative hypothesis should we use to conduct a statistical test?

  b. Using the information in the results section of this story to conduct a test of proportion using the data from 1996.

  c. Part of the model being tested under the null hypothesis is that the data are assumed to be a series of independent trials. Are each of the trials in this study independent? Explain.

Online Problems for Chapters 22 - 23

Open the Web site: www.whfreeman.com/applet/pvalue.html
This Web page has an applet that calculates probability values (p-values) and simulates random samples under a null or alternative hypothesis. On the applet you will see the outline of a normal distribution that represents the sampling distribution of the sample mean. You will also see several boxes that allow you to specify a null hypothesis, sample size, and value of the standard deviation. This applet can be used in two ways – to calculate the p-value from a known sample and to simulate the many samples when the population is known.

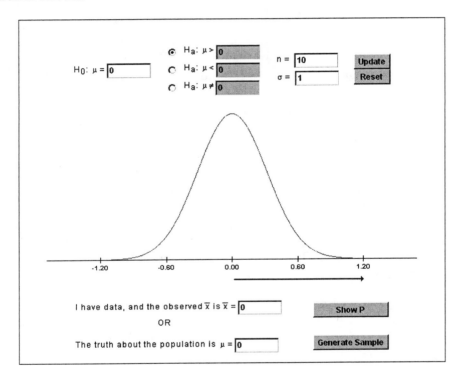

206. Consider problem 195 above. Use this applet to calculate the p-value for this test.

   a. Set the null hypothesis to 2000 and the alternative hypothesis to greater than 2000. Notice that when you put the value in the null hypothesis, the alternative hypothesis automatically changes to reflect that value. This is because the alternative should always match the null hypothesis.

   b. Set the sample size and standard deviation and click "update." You will notice that the normal curve will change to reflect this setting.

   c. Enter the sample mean 2240 in the box for the observed $\bar{x}$. Click "Show P." Does this p-value match the p-value you calculated in problem 195?

207. Consider again the situation of the previous problem. We can use the applet to
· generate random samples to see how often results would be as extreme as those we
observed. Do this by again setting the null hypothesis to 2000, sample size to 25, and
standard deviation to 600.

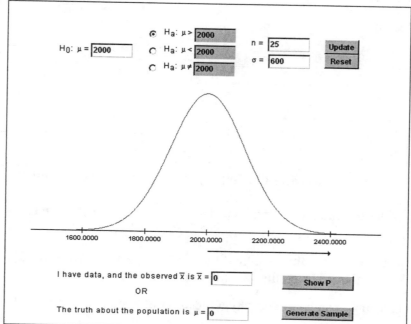

a. We will begin by seeing what will happen when the null hypothesis is true. Set
the box marked "The truth about the population is μ = " 2000. Click "Generate
Sample." A sample mean and p-value will appear. Repeat this process 25 times
and record the means and p-values here.

Means:

_____ _____ _____ _____ _____ _____ _____ _____
_____ _____ _____ _____ _____ _____ _____ _____
_____ _____ _____ _____ _____ _____ _____

P-values:

_____ _____ _____ _____ _____ _____ _____ _____
_____ _____ _____ _____ _____ _____ _____ _____
_____ _____ _____ _____ _____ _____ _____

b. In how many of these 25 times did you see a mean of 2240 (what was observed in
actual sample) or more?

c. In how many of these cases would you have rejected the null hypothesis? That is,
in how many of the cases did you see a p-value less than 0.05?

d. Next let's assume that the alternative hypothesis is true and that this population really has a mean of 2200. Leave all other values the same but set "The truth about the population is $\mu =$" 2200. Repeat the procedure 25 times and record the values here.

Means:

___ ___ ___ ___ ___ ___ ___ ___ ___
___ ___ ___ ___ ___ ___ ___ ___ ___
___ ___ ___ ___ ___ ___ ___

P-values:

___ ___ ___ ___ ___ ___ ___ ___ ___
___ ___ ___ ___ ___ ___ ___ ___ ___
___ ___ ___ ___ ___ ___ ___

e. In how many of these cases did you observe a p-value that was less than 0.05?

208. Again consider the situation in problem 14. Assume that the observed sample mean was 1850.

a. Use the applet to find the p-value if the observed sample mean was 1850. Explain why this p-value takes on the value that it does.

b. What conclusion about the null hypothesis would you make in this situation?

144

Additional Problems for Chapter 24

209. The cross-classified data table below gives information on whether a convicted murderer gets the death penalty broken down by the race of the victim and the race of the defendant.

| | **White Defendant** | | **Black Defendant** | |
Race of victim	Death sentence	No death sentence	Death sentence	No death sentence
White	19	132	11	52
Black	0	9	6	97
Total	19	141	17	149

   a. What percentage of White defendants received the death penalty? What percentage of Black defendants received the death penalty?

   b. Compute the rate of death sentences for the races of defendants separately for each race of victim.

   c. Explain (in statistical/quantitative terms) any discrepancies you see between your responses to parts a and b.

210. America West Airlines and Alaska Airlines compete for passengers flying into five major cities in the western United States. The cross-classified data table below gives the on-time performance for the two airlines, in June 1991, taken from their monthly report to the FAA.

| | **Alaska Airlines** | | **America West Airlines** | |
Destination	Flights on time	Flights not on time	Flights on time	Flights not on time
Los Angeles	497	62	694	117
Phoenix	221	12	4840	415
San Diego	212	20	383	65
San Francisco	503	102	320	129
Seattle	1841	305	201	61
Total for 5 cities	3274	501	6438	787

   a. What percentage of America West Airlines flights were on time in Phoenix? What percentage of the on-time America West Airlines flights were in Phoenix?

   b. Overall, for these five cities, Alaska Airlines was on time for 86.7% of their flights whereas America West airlines was on time for 89.1% of their flights. America West ran several advertisements during the summer of 1992 citing the FAA data and claiming to have the best on-time record. Why isn't this claim justified by the data in the table above? Explain briefly.

211. A study is conducted to compare the quality of the neurosurgical staff at each of three hospitals in the same city. The researchers examine the records of patients who were operated on for brain tumors during the period between 1989 and 1993. For each patient they recorded the tumor grade and whether the patient was still alive one year after surgery. The results are given in the table below.

Hospital	Patients with High-grade tumors		Patients with low-grade tumors	
	Still alive after 1 year	Died within 1 year	Still alive after 1 year	Died within 1 year
University hospital	40	40	25	5
County hospital	20	30	80	30
Private hospital	15	20	60	20
Total	75	90	165	55

a. What percentage of patients with high-grade tumors survive for at least one year after surgery? What percentage of the patients at County Hospital with high-grade tumors survive at least one year after surgery?

b. In a summary of the report's findings, the authors write:
"The neurosurgical staff at the private hospital seem to have the best record, with 65.2% of patients surviving at least one year; the county hospital had the second best record with a 62.5% survival rate; the university hospital had the worst record with only a 59.1% survival rate. Patients having brain tumor surgery should be advised to stay away from university hospital."

Explain why the conclusion of this report is not justified. Based on the data in the table, which of the three hospitals appears to have the best record?

212. Zocor is a drug manufactured by Merck and Co. that is meant to reduce the level of LDL Cholesterol in the blood, which in turn should reduce a number of problems like heart attacks and strokes. In a study, "20536 UK adults (aged 40 – 80 years) with coronary disease, other occlusive arterial disease, or diabetes were randomly allocated to receive 40 mg Zocor daily or a placebo." The Merck Web site further reports that of the 10,269 subjects on Zocor, 898 had a "major coronary event" (heart failure); while of the 10,267 on the placebo, 1212 subjects had a major coronary event. Is there a significant relationship between the treatment (Zocor vs. placebo) and if the subject had a major coronary event? Conduct a chi-square test for independence to determine this. Be sure to specify a null and alternative hypothesis, test statistic, p-value, and conclusion.

213. Large trees growing near power lines can cause substantial damage during storms from falling limbs. Researchers are developing two chemical treatments to stunt and slow the growth of trees. However, if the treatment is too severe the tree will die. In an experiment on 275 sycamore trees, researchers applied the two chemical treatments (Chemical 1 and Chemical 2).

Treatment	Survival	
	Lived	Died
Chemical 1	94	41
Chemical 2	127	13
Total	221	54

Conduct an appropriate test of hypothesis to see if there is a significant difference between the two chemicals. Be sure to specify a null and alternative hypothesis, test statistic, p-value, and conclusion.

214. A study was conducted to evaluate the relative efficacy of supplementation with calcium versus Calcitriol in the treatment of postmenopausal osteoporosis. (Calcitriol is an agent that has the ability to increase gastrointestinal absorption of calcium.) A number of patients withdrew from this study prematurely due to the adverse effects of treatment, which include thirst, skin problems, and neurological symptoms. Conduct an appropriate test of hypothesis to see if a patient withdrawal is related to the treatment the subject was receiving.

Treatment	Withdrawal	
	Yes	No
Calcitriol	27	287
Calcium	20	288
Total	47	575

Applet Activities

The Mean and Median applet

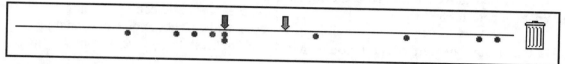

Address: www.whfreeman.com/applet/mean.html

Directions: Use your mouse to click below the line and make red points creating a dot plot (points that fall at the same place along the number line stack up to give the appearance of a histogram). Above the line, the applet shows the mean represented by a green arrow and the median represented by a red arrow. When the mean and median are the same, the overlapping arrows form one yellow arrow.

Contemplate:

1. If you create just one point, where will the mean appear? How about the median?

2. If you create two points, where will the mean appear? How about the median?

3. If you put two points at the far left side of the line and one point at the far right (near the trash can), where will the median arrow appear? How about the mean arrow?

Verify: Use the applet to test your answers to the three questions above. Were the mean and median the same when you thought they would be? Did the mean and median fall exactly where you expected in each of the three cases?

Applet tip: Click on the trash to clear the points.

149

Experiment: Create a histogram of four points in a symmetric pattern at the left side of the line. Notice how the mean and median are overlapping and right at the point of symmetry. Now add a fifth point just to the right of the others. Are the mean and median still overlapping? Are they close? Now drag this point to the right and watch the behavior of the mean and the median. What happens to the mean as you move the point to the right? What happens to the median?

Applet tip: You can grab a single point and drag it . You can even drag it into the trash to remove it from the plot.

Do this again, but start with ten points in a symmetric pattern at the left side. As you move an 11th point to the right how does the mean change? How does the median change? How do these changes compare with what happened when you started with just five points?

The Normal Density Curve applet

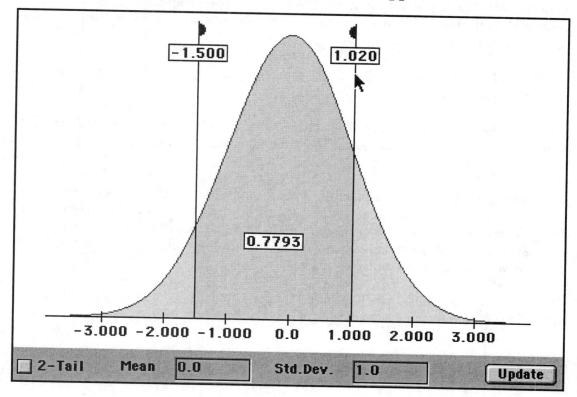

Address: www.whfreeman.com/applet/normal.html

Directions: Drag the green flags to have the applet compute the part of the normal curve that is in the bright yellow shaded region you create. When the "2-tail" box is checked the applet calculates symmetric areas around the mean.

Contemplate:

1. If you were to put one flag at the extreme left of the curve and the second flag right in the middle, what would be the proportion reported by the applet?

2. If you were to place the two flags exactly one standard deviation on either side of the mean, what would the applet say is the area between them?

3. About 99.7% of the area under the normal density lies within three standard deviations of the mean. Does this mean that about 99.7%/2 = 49.85% will lie within one and a half standard deviations?

Verify: Use the applet to test your answers to the three questions above. Did the shaded proportion of the curve behave as you thought it would in each case?

Applet tip: The direction that the green flags point is the side of the flag poles that is shaded.

Experiment: Try comparing the values given by the applet with those given by Table B in the back of the text to see that they are the same (examine at least one positive standard score and one negative standard score).

Use the applet to find the quartiles of the normal distribution when the mean is 0 and the standard deviation is 1.

The 68-95-99.7 rule tells us that the middle 95% of values fall within about 2 standard deviations of the mean. Use the applet to get a more precise answer for where the middle 95% of values fall.

Make a small table of the areas under the normal curve within ±0.5, ±1.0, ±1.5, and ±2 of the mean. Next, change the standard deviation from 1 to 2 and note how the numbers along the horizontal axis have changed. Again make a table of the areas within ±0.5, ±1.0, ±1.5, and ±2 of the mean. Do any of the areas in this second table match some of the areas you have in your first table? Explain.

Solve any of the problems from 13.18 to 13.27 in the text using both the applet and again using Table B. Do you find any advantage(s) to using the applet? Explain.

The Correlation and Regression applet

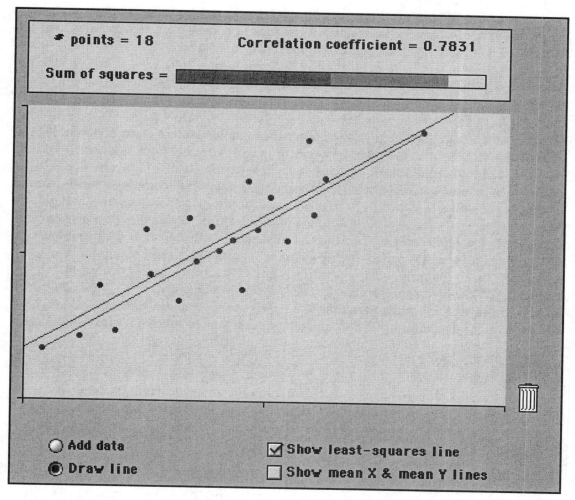

Address: www.whfreeman.com/applet/corr.html

Directions: Use your mouse to click on the blank plot to make red points creating a scatterplot. Above the plot, the applet shows the correlation. Below the plot are options that can be checked if you wish to draw the regression line on the plot, see the average value of x and y, or even add your own line to the plot to see how well your line competes with the regression line.

Contemplate:

1. If you create just two points, what would the correlation be?

2. If you put three points in a tight, positively associated linear pattern in the upper-left-hand corner of the plot, what will the correlation be? If you then add a fourth point at the

lower-right-hand corner (near the trash can), what will happen to the correlation? Where would the regression line fall?

3. If you put 10 points in the center of the plot in a U-shaped pattern, what will the correlation be? Where will the regression line fall?

Verify: Use the applet to test your answers to the three questions above. Did the correlation and regression line behave as you thought they would?

Experiment: Create a scatterplot of 10 points in a random pattern at the bottom left-hand side of the plot. Move the points around to make the correlation as close to zero as possible. Have the applet show the least-squares regression line. What is its slope? Now add an 11th point at the upper-right-hand corner and watch the behavior of the correlation and the regression line.

Next, clear the scatterplot by clicking on the trash and draw a new group of points in a roughly linear pattern with a positive association. Grab one point and drag it around the scatterplot while watching the

> **Applet tip:** Watch the "sum of squares" bar to see how well your line competes against the best line. The green part of the bar shows how much worse you're doing.

behavior of the correlation. Where does the placement of the point create the largest correlation? Where does the point create the smallest correlation? Does the correlation ever become negative?

Clear the scatterplot again and create a cloud of points that has about 25 points in a weak linear association (e.g., a correlation of about 0.5 or –0.5). Draw a line by hand that goes through the center of the cloud trying to also capture the slope of your cloud of points. Now have the applet add the regression line to the plot. How does your line compare? Did you make the slope steeper than the regression line? Does your line cross the regression line near the center of the cloud? Focus on the points that lie farthest to the right on the plot (say the points with the highest 10 to 20% of the x values). Does the regression line seem to be going through the average y-value for these points with the high x-values? Does your line behave in this way or did it tend to fall too high or too low to capture the average y-value?

Try moving your own line around the plot while watching the green part of the "sum of squares" bar. Can you make the green part disappear without putting your line right on top of the regression line?

> **Applet tip:** You can clear your line from the plot by dragging one of its end-points into the trash.

Explain. Place your line right on top of the regression line, and remove the regression line from the picture. Now switch back to the "add data" mode and put an additional single point on the plot. Grab this point and move it around the scatterplot while watching the "sum of squares" bar. Where does the placement of the point create the largest green area (indicating your line is no longer close to the regression line)? Can you find several places to put the point where the green part of the bar disappears? Explain.

The Simple Random Sample applet

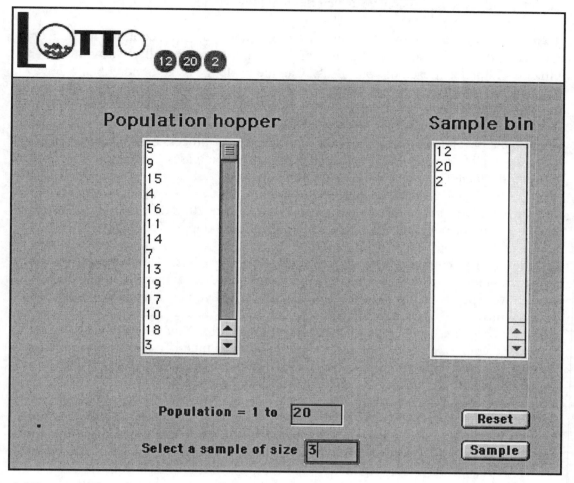

Address: www.whfreeman.com/applet/srs.html

Directions: Enter the number of lotto balls you want in the population (500 maximum) and click the reset button to fill the population hopper. Enter the sample size and click the sample button to put the sampled lotto balls in the sample bin and remove them from the population hopper.

Contemplate:

1. If you were to sample three balls from a population of five, which is more likely to happen: a) getting the sample 1, 2, and 3 in order; or b) getting the sample 3, 1, and 5 in order ?

2. If you were to repeatedly draw samples of size three from a population of size five, how often would the largest of the three numbers be picked first?

> **Applet tip:** You can use the applet to put the population in a random order by making the sample size and the population size the same.

155

Verify: Test your answers to the two questions above by making repeated use of the applet. Did the samples behave as you thought they would?

Experiment: Explain how you would use the applet to a) flip a coin, b) shuffle a brand new deck of cards, and c) pick a jury of 12 people from a small town with 500 adult citizens. Implement each of your ideas. Can the applet be used in more than one way to implement each of these tasks? Explain.

Try completing exercises 2.7, 2.8, 2.9, 2.11, or 2.12 from the text using the applet instead of the table of random digits. Does the applet make the process easier or more difficult? Does the applet provide a safeguard against any potential biases that you feel might occur in using the table for drawing a sample? Explain.

Draw a sample of seven balls from a population of size 20. After the sample is drawn, what is left in the population hopper? Is this a random sample? Explain.

(The following should be done by every student in the class.) Use the applet to pick a simple random sample of five students from your class. How many men and how many women were picked in your sample? Which gender did you have more of in your sample than would be expected from the composition of the whole class? Does this show that your sample was biased? How many of your classmates' samples contained more men than expected and how many of your classmates' samples contained more women than expected? Find the average percentage of men found in all of your classmates' samples put together. How does this compare with the actual percentage of men in the class? What properties of the random sample are illustrated by the results from your class? Explain.

> **Applet tip:** You can sample up to 40 balls at a time. To make the sample size larger than 40, you can string two samples together.

156

The Probability applet

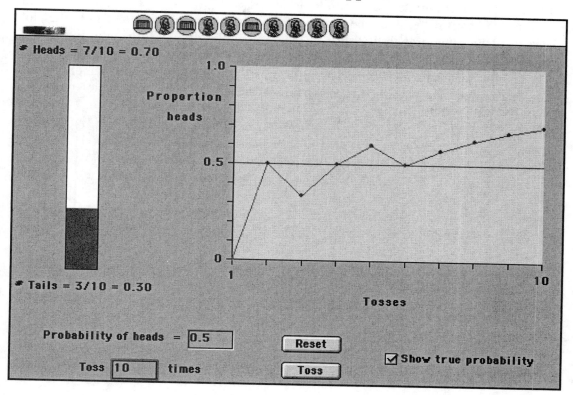

Address: www.whfreeman.com/applet/prob.html

Directions: Click the "toss" button to have the applet toss a coin the number of times you specify. Set the probability that the coin lands heads by entering a new value and clicking the reset button.

Contemplate:

1. If you toss the coin just once, what are the possible values for the proportion of heads?

2. If you toss the coin 400 times with a probability of 0.5, which is more likely to happen: a) getting exactly 200 heads, or b) getting within a few percent of 200 heads?

3. What will the proportion of heads plot look like as you accumulate more and more tosses with a probability of 0.7 for heads?

Verify: Use the applet to test your answers to the three questions above. Did the proportion of heads behave as you predicted?

Applet tip: You can toss the coin a maximum of 40 times at once. To make longer sequences, toss again without resetting.

Experiment: Set the probability of heads at 0.3 and have the applet repeatedly toss the coin until you accumulate many hundreds of tosses. How does the behavior of the proportion of heads differ at the beginning of the plot compared with the end of the plot. Check the "show the probability" box to have the applet draw a horizontal line at 0.3. How does the behavior of the plot of the proportion of heads relate to this line? Change the probability of heads to 0.7 and have the applet toss the coin several hundred times again. How does the plot of the proportion of heads compare to the plot when the probability was 0.3? In what ways is it similar? It what ways does it differ?

The Expected Value applet

Address: www.whfreeman.com/applet/lln.html

Directions: Use the "Fewer dice" and "More dice" buttons to fix the number of dice to be rolled with each trial. Next, enter the number of rolls (100 maximum) and then click the "Roll dice" button to have the applet roll the dice and plot the accumulating average of the roll totals.

Contemplate:

1. When you just roll one die, what values can you get? Are they all equally likely?

2. When you roll 10 dice, what values can you get for the roll total? Are they all equally likely?

3. Which process is likely to get close to its mean value in fewer trials: a) the accumulating averages of rolling one die at a time, or b) the accumulating averages of rolling 10 dice at a time?

159

Verify: Test your answers to the three questions above through repeated use of the applet. Did the individual row totals and the accumulating averages behave as you thought they would?

Experiment: Have the applet roll one die 100 times and observe the behavior of the accumulating average plot. How far from the long run mean was the average after 10 rolls? after 20 rolls? after 50 rolls? .after 100 rolls? Explain the pattern. Reset the applet and roll 10 die 100 times. Describe the pattern you see in the accumulating average. How is this pattern similar to the pattern for rolling one die? How does it differ?

The average of the numbers from 1 to 6 is $(1+2+3+4+5+6)/6 = 3.5$. Set the applet to roll one die and check the "Show mean" box to see that the expected value for rolling a single die is 3.5. Now alternately click the "More dice" button and the "Roll dice" button while you watch the behavior of the expected value for the sum of the rolls as more dice are added. What is the pattern?

What is the smallest total you can get with one roll and what is the largest total you can get? (*hint*: the answer depends on how many dice you are rolling.) Regardless of the number of dice rolled, the vertical axis on the accumulating average plot is marked off in 10ths of the whole range of possible values. Have the applet roll one die 20 times and count how many times the individual roll total falls within the first tick mark to either side of the mean. Have the applet make 20 replications of rolling two dice at a time, then three dice at a time, then four, etc..., each time checking how many of the individual roll totals fall within one mark of the mean value. Describe and explain the pattern you observe.

The Confidence Intervals applet

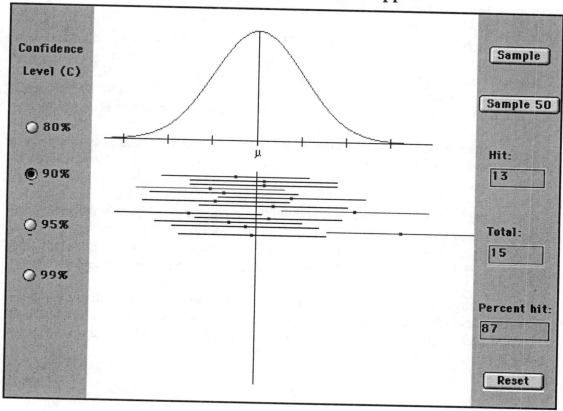

Address: www.whfreeman.com/applet/CI.html

Directions: Click the "sample" button to have the applet take a random sample and construct a confidence interval. Click the "Sample 50" button to have the applet take 50 different random samples and make 50 different confidence intervals from the resulting data.

Contemplate:

1. If you just make one 90% confidence interval, will it necessarily capture the parameter value?

2. If you make 10 different 90% confidence intervals, how many of them do you think will capture the parameter?

3. If you make one thousand 90% confidence intervals, how many of them do you think will capture the parameter?

> **Applet tip:** The cumulative totals are cleared when you click the "reset" button. When you change the confidence level, they pertain only to the displayed intervals.

Verify: Test your answers to the three questions above through repeated use of the applet. How far did the results vary from what you expected in questions 2 and 3? How

161

far did the results vary from what you expected on a percentage basis? Comment on the principles this illustrates.

Experiment: Set the confidence level to 80% and sample 50 confidence intervals. How many of these intervals fail to capture the parameter (they are drawn in red)? Increase the confidence level in succession to 90%, then to 95%, and then to 99% while watching the changing behavior of the 50 intervals. What happens to the lengths of the intervals? What happens to the number of intervals that fail to capture the parameter?

Have the applet make 50 new confidence intervals. The dots in the center of the intervals mark the sample mean. Are the intervals produced by the applet symmetric around the sample mean?

The tick marks on the horizontal axis below the normal curve are each separated by one standard deviation of the sample mean ($\sigma/\sqrt{n}$). According to the 68-95-99.7 rule, how many of the 50 sample means would you expect to fall within one standard deviation of the parameter? How many of the 50 samples did fall within one standard deviation?

How is the margin of error related to the length of the interval? Explain. Examine the 50 intervals at the 95% confidence level. What is the typical margin of error for these intervals measured in units of the standard deviation of the mean? Explain how this is related to the normal curve.

The Test of Significance applet

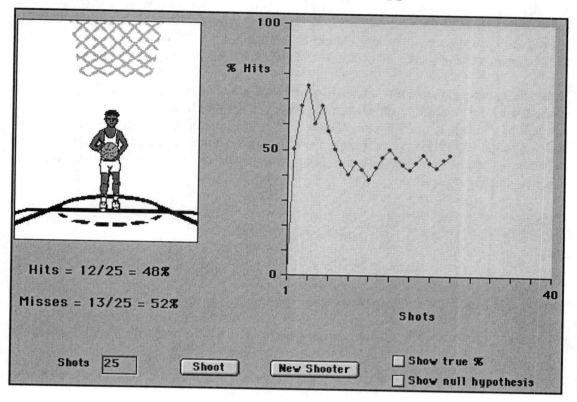

Hits = 12/25 = 48%

Misses = 13/25 = 52%

Shots `25`   **Shoot**   **New Shooter**   ☐ Show true %
☐ Show null hypothesis

Address: www.whfreeman.com/applet/hyptest.html

Directions: Enter the number of shots (100 maximum) and click the "Shoot" button to gather data on the free-throwing ability of the shooter. Click the "New Shooter" button to change the true value of p = the probability that the shooter makes a basket. The plot shows the accumulating percentage of shots made ($\hat{p}$) and can include horizontal lines at the null hypothesis (0.8) and at the true value of p for the current shooter if those boxes are checked.

Contemplate:

1. Draw a picture that shows what you think the plot of the accumulating percentage of shots made will look like when the null hypothesis is true.

2. Draw a picture that shows what you think the plot of the accumulating percentage of shots made will look like when the null hypothesis is not true.

> **Applet tip:** The null hypothesis is always $p = 0.8$, and the alternative hypothesis is $p < 0.8$ because the applet never fixes p at a value higher than 0.8.

Verify: Use the applet to test your two answers above. Did the plots look similar to your drawings? What key aspect(s) of the plots did you capture? What aspect(s) differed?

Experiment: Use the applet to have the shooter make five shots. Based on this data, can you decide if the shooter can hit 80% of his shots overall? Explain why or why not. Add another five shots to the total. Can you decide at this point? Continue to add five shots at a time until you are confident that you know whether the null hypothesis is true or false. How many shots did it take? Check the "Show true %" box to see the true value of p.

Remove the true value of p from the plot and pick a new shooter. Repeat the process of using the applet to add five shots at a time until you feel confident that you know whether the null hypothesis is true or false. Did it take more shots or less shots this time? View the true value of p. Did it take longer for you to determine whether the null hypothesis was true or false when p was closer to 0.8 or when it was farther from 0.8? Explain why this happened.

Appendix A: Using Data Desk®

Throughout the labs and activities in this manual, it is important to use the computer as a tool for investigating the theory that you read in the text and learn in lecture. Many statistical analysis packages offer the features needed to investigate statistical concepts. This activity will help you pick up some of the basics of using Data Desk. A typical Data Desk window looks similar to any window on your computer's desktop.

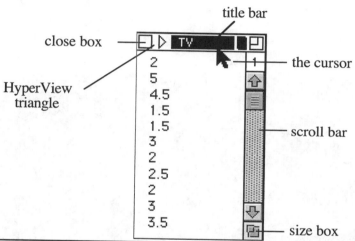

Term	What it means
choose	click on a menu – holding the mouse button until you pick the desired option
close	click in the close box to make the window disappear
move	click on the title bar and drag the window to a new location
open	move the cursor to an icon and double-click
resize	click in the size box and drag to make the window bigger or smaller
scroll	click in the scroll bar or on the adjoining up or down arrows
select	move the cursor to an icon and click once (the icon is then highlighted)

Now we are ready to try the Data Desk software. Ask your instructor for a demonstration if something is unclear.

Have your instructor show you where to find the data files you'll be using in this class and open the "Cereals" data file.

A picture showing what your desktop should look like (using version 6 of Data Desk) is given below. Data Desk's File Cabinet ( ) contains folders holding the data ( ) and the results ( ) of any statistical procedures that you carry out. The window labeled "Cereals" contains icons ( ) that represent the variables in the Cereals data as well as a reference scratchpad ( ) that gives a description of the source of the data.

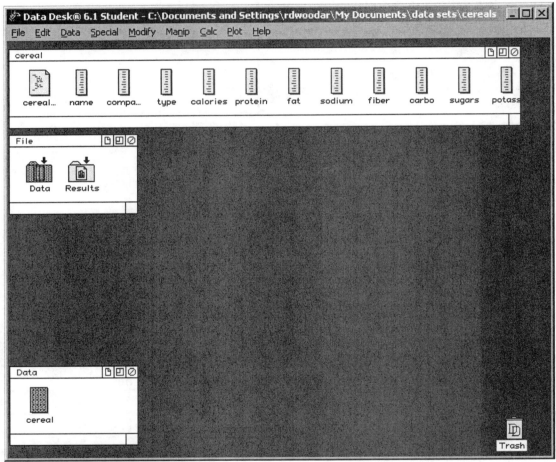

Open the 'Reference' scratchpad.

1. What is the source of the "Cereals" data?

The variables 'company' and 'calories' describe the manufacturer and the amount of calories of these cereals. Move the reference window to the lower part of the screen and open the icons for the 'manufacturer' and 'calories' variables. Scroll down the list of manufacturers and notice how the two variables are linked, and scroll together.

2. Examine the data and find Kellogg's All Bran. How many calories does this cereal have?

Close the 'Reference' and the two variable icon windows. Most statistical analyses are carried out in Data Desk by selecting the variables you wish to work with and then choosing the analysis you want to do from the appropriate menu item. Select the 'company' variable (the icon will now appear shaded) and choose **Plot>Pie Charts**. A window will appear with a pie chart showing the fraction of cereals in this list that are from each of the manufacturers. Move the pie chart window to the lower-right-hand part of the screen.

3. Which company produced the most cereals in this data set?

Locate the small HyperView triangle in the upper left of the pie chart window (▢ ▷ ▇).

The HyperView menus give you additional suggestions for analysis. Click on the triangle to see the list of choices. Before releasing the mouse button, choose "Frequencies of company" (it will be highlighted) and then release the button. Your desktop should look something like the picture on the next page.

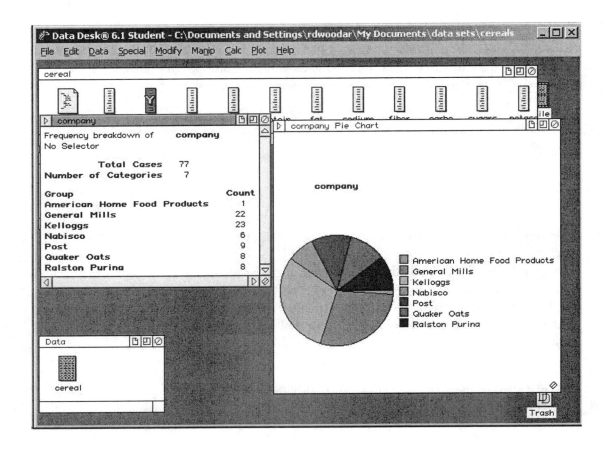

4. *How many cereals are studied in this data set?*

5. *How many of these cereals are from Kellogg's?*

To add percentages to your frequency table, locate **Frequency Options** in the HyperView triangle menu and check off the percent option.

6. *What percent of the cereals are from Nabisco?*

Next we will learn how to obtain some numerical summaries of a variable. Data Desk allows you to choose what type of summaries you want. To do this, click on **Calc>Calculation Options>Select Summary Statistics**.

A window will appear with a set of choices for summary statistics that you may select. Each choice is preceded by a small box. When you click in an empty box to make a check appear, the corresponding data summary will be computed. Clicking in a box that is already checked will make it disappear. You should select the following options: "Mean," "Total cases," "Non Numeric," "Min," and "Max." The Summary Options window with these options and selected looks like this:

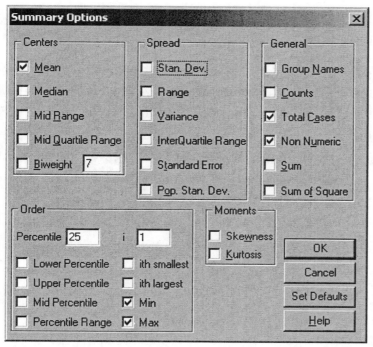

Click on the "OK" button to save your selections and close the window.

Select the 'calories' variable and click **Calc>Summaries>Reports.** Data Desk will calculate the summary statistics that you have requested for calories variable.

8. What is the mean (or average) number of calories for cereals in this data set?

9. What is the lowest number of calories for any cereal in this data set?

10. What is the highest value?

Make a bar chart of the 'type' variable. To do this, click on the variable and then click **Plot>Bar Charts**.

11. Are there more hot or cold cereals in this data set?

Instead of working with a data file that has already been created, we often need to enter data into a new file on our own. To do this, you can open a new data file {**New Datafile** in **File** menu} and create a variable by choosing **Blank Variable** in the **New** submenu under the **Data** menu. Data Desk will display a dialog requesting a name for the new variable. To enter data, type values one row at a time and press the Enter key. When you need to create more than one, be sure that you enter data one case at a time, moving along a row using the Tab button until it is complete (then press Enter to move to the next row).

12. Create a new variable. Using your mobile phone look through the 10 most recent outgoing calls. Enter the time for each of these calls into your new variable and calculate the average amount of time for these calls.

Summary

In Data Desk, variables are represented by icons. Highlight the variables you want to work with and then use the menu items to perform the desired operations. When you want to use two variables, the first one you click on will be the *y*-variable. Hold down the shift key to select *x*-variable(s).

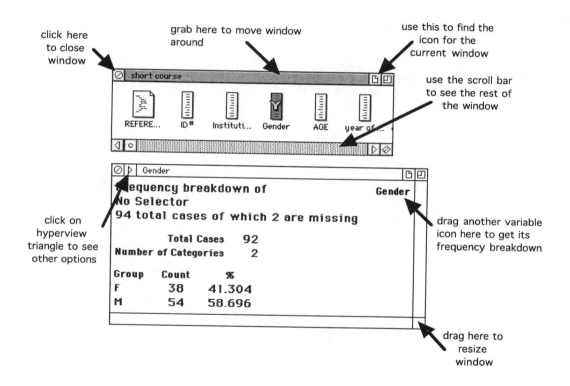

click here to close window

grab here to move window around

use this to find the icon for the current window

use the scroll bar to see the rest of the window

click on hyperview triangle to see other options

drag another variable icon here to get its frequency breakdown

drag here to resize window

short course

REFERE... ID# Instituti... Gender AGE year of...

Gender

Frequency breakdown of **Gender**
No Selector
94 total cases of which 2 are missing

| | Total Cases | 92 |
| Number of Categories | | 2 |

Group	Count	%
F	38	41.304
M	54	58.696

Throughout the activities and problems you will find in this manual we make use of statistical software to make calculations and creation of graphics easier. One program that is very useful for statistical analysis is CrunchIt!, which runs through a browser window. It is available for free at http://bcs.whfreeman.com/crunchit/scc using the access code that came with this lab manual. The program is also available under the name StatCrunch through the Web site www.statcrunch.com (see screenshot below). In this appendix we will introduce you to the basic use of CrunchIt.

Using CrunchIt!

Go to the Web site http://bcs.whfreeman.com/crunchit/scc and enter your access code to get into CrunchIt. A new Web window will appear that resembles a spreadsheet. Across the top of this window you will see several menus that allow users to work with data. For this demonstration we will make use of a data set that includes the nutritional information of many popular brands of cereals. You can read more about the data in the EESEE story *Nutrition and Breakfast Cereals*. This data set is also included in the Sample Data sets within CrunchIt.

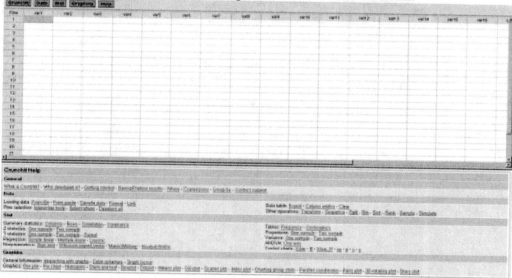

1. To open a data set simply click on **DATA>Load Data>Sample Data.** CrunchIt can also load data from files housed on Web sites or on your hard drive. For now we will make use of the cdereal data. Before opening the data set, take a moment to examine the "info" link that defines the variables included in this data set.

CrunchIt Sample Data Sets:

Click on the name of the data set to load it into CrunchIt.
Click on the info link for a description of the data set.
Directories containing other data sets are shown in bold.

- **Journal of Statistics Education Data Sets**
- Body measurements of sparrows - info - 1K
- CEO salaries - info - 681b
- Cereal data - info - 11K
- CPS wage data from 1985 - info - 15K
- Cuckoo - info - 1019b
- Fortune billionaires in 1992 - info - 7K
- Highway deaths - info - 537b
- Home prices in Albuquerque - info - 3K

You will see that this data set includes information on breakfast cereals from many different manufacturers, which are abbreviated with the first letter of their names. Next, click on the cereals data to open the data set.

2. *Examine the data and find Kellogg's All Bran. How many calories does this cereal have?*

Creating graphics in CrunchIt is relatively easy. For categorical variables such as the manufacturer (mfr) you can create a pie chart. To create a pie chart select **Graphics>Pie chart> With Data**. You then should choose the appropriate variable or variables from the *Select columns* window. Then click "Create Graph!"

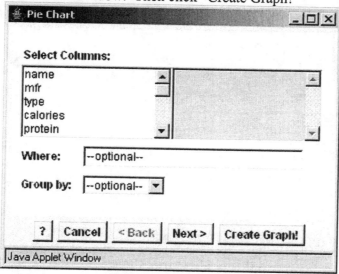

3. *Create a pie chart for manufacturer and examine the results. Which company produced the most cereals in this data set?*

Any output in CrunchIt can be saved or printed by using **Options>Export(Save/Copy/Print).**

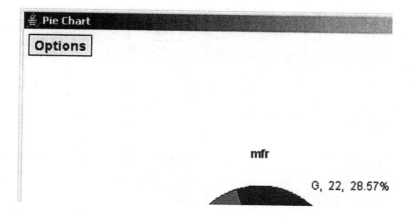

Producing summary statistics for each variable is also easy. For numeric variables such as the amount of protein or calories in the cereals, click on **STAT>Summary Stats> Columns,** in the main window.

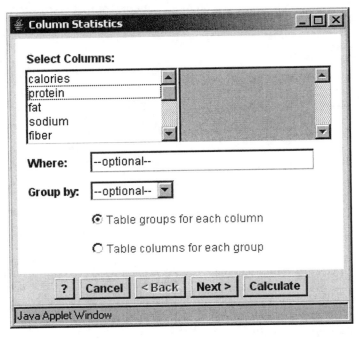

4. *Calculate the average amount of protein for all cereals.*

For categorical variables such as manufacturer or type of cereal you can get percentages by using **STAT>Tables>Frequency.**

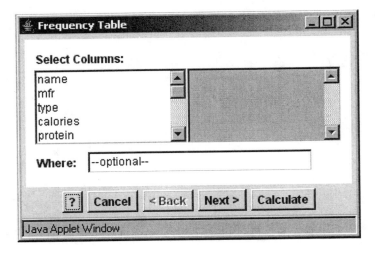

5. *Create a table of the manufacturers. You will see that it provides both the number in each category (frequency) and the proportion in each category (relative frequency). From this table what proportion of the cereals was produced by Post?*

Appendix C: Using Minitab®

This activity will help you pick up some of the basics of using Minitab to calculate summary statistics, produce graphical displays, and investigate statistical concepts. Minitab is specifically designed to be easy to use and to include many of the standard statistical procedures that are used in the Lab and Activities Supplement.

Basic layout of the Minitab screen

After opening Minitab you will see a screen that has three distinct parts. The Menu Bar across the top of the screen has many of the standard menus found in other programs, as well as some that are specific to this program. There is also a Session Window in which the results of your statistical procedures will appear.

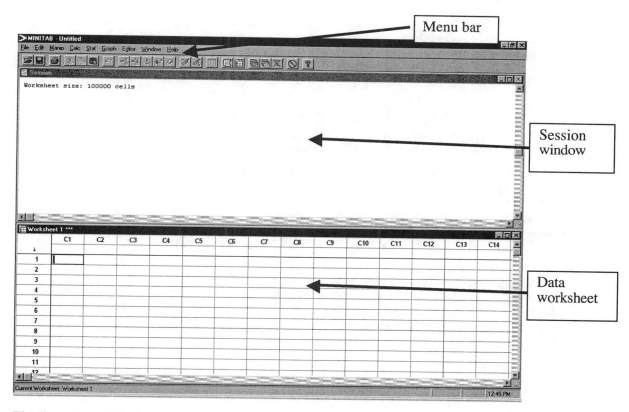

Finally, a Data Window fills the remainder of the screen. The data window allows you to enter or change data. The data window is arranged in a series of columns and rows much like a spreadsheet program. Columns are labeled C1, C2, C3, etc., while rows are numbered 1, 2, 3 etc. The first row of each column is reserved for names, and is not numbered. Data can be entered in the numbered rows.

Opening data sets

While you can enter data sets of your own creation you may wish to use a data set that has already been created. To do this you will need to open the data file. Opening data files in Minitab works like opening a file in most programs. From the **File** menu choose **Open Worksheet.** If the data set you are opening is not a Minitab worksheet you may need to choose the appropriate file type. Minitab can import files from a variety of popular programs such as Microsoft Excel or from simple text files.

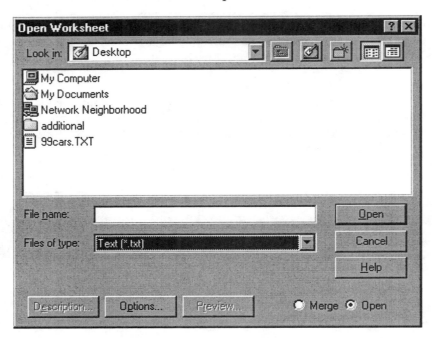

For this example we will use the "99Cars" data. Your instructor can help you find this file on your lab computer or you can download it from the www.whfreeman.com/scc Web site. The "99 Cars" data has information on different cars from the 1999 model year. These cars were subjected to testing by the EPA, and various characteristics were recorded. After opening the data set, you will see that each of the individual makes of car is in a different row of the data window and each variable is in a different column.

Creating Graphics

Often we would like to summarize a variable in a data set through the use of a graphic. Minitab makes the creation of professional quality graphics easy. For instance you can create a pie chart from the variable that gives the car's type of drive. Clicking on **Pie Charts** under the **Graph** menu will open a dialog box.

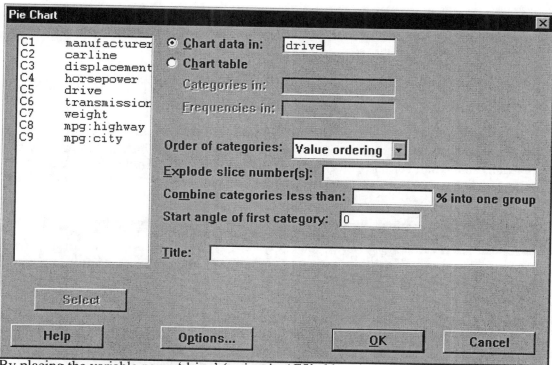

By placing the variable name 'drive' (or just its 'C5' abbreviation) in the **"Chart data in"** box you can create a pie chart of the type of drive variable.

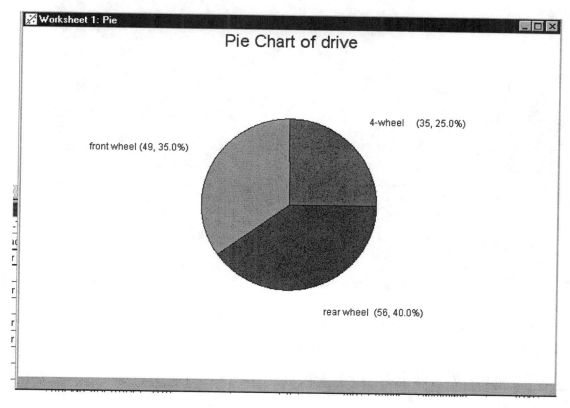

1. Is there approximately an equal proportion of front-wheel, rear wheel, and 4-wheel-drive vehicles in the EPA data set? Explain briefly.

Minitab can also produce numeric summaries of variables. For instance we may wish a simple tally of the number and percentage of each type of drive. In the **Stat** menu choose **Tables** and **Tally.** In the Tally Window, choose the name of the variable you would like to summarize and check off the types of summaries you would like.

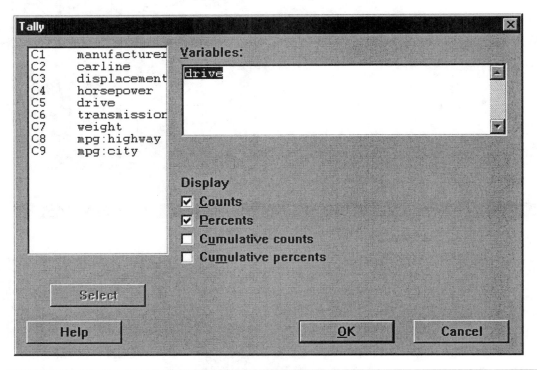

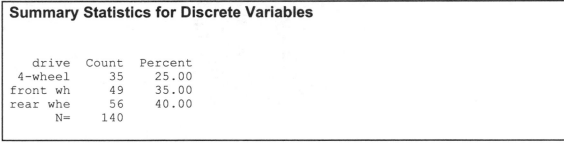

Summary Statistics for Discrete Variables

drive	Count	Percent
4-wheel	35	25.00
front wh	49	35.00
rear whe	56	40.00
N=	140	

2. How many car models are studied in this data set?

3. How many of the models are 4-wheel-drive vehicles?

4. What percent of the cars in this list are front-wheel drive?

Some variables like the city mileage of the cars (mpg:city) have many possible values and creating simple counts and percentages would not be informative. Instead we often summarize these variables with summary statistics like averages or medians. These summary statistics are easily found by choosing **Basic Statistics** and **Descriptive Statistics** from the **Stat** menu.

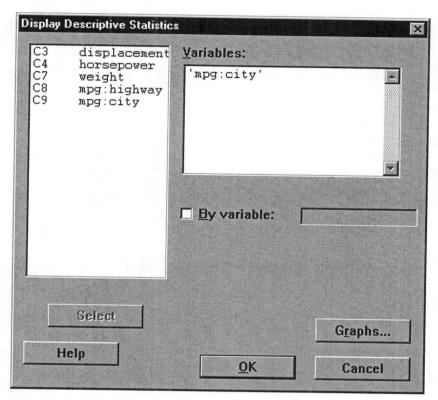

5. What is the mean (or average) city mpg value for the cars in this data set?

6. What is the lowest mpg value?

7. What is the highest mpg value?

Review

Make a Dotplot of the displacement variable. *(hint:* look under the **Graph** menu.)

8. The Mercedes CLK430 has a 260 cubic inch engine. Are there more cars with bigger engines than this Mercedes or are there more cars with smaller engines? (Explain how you can answer this from looking at the dotplot.)

Create a summary of the highway gas mileage.

9. What is the median highway gas mileage for the cars in this data set?

New worksheets.

Instead of working with a data set that has already been created, we often enter data into a new file (called a worksheet by Minitab) on our own. This can be done from the **File** menu by choosing **New** and **Minitab Worksheet.** This will open a new worksheet on which you may enter additional variables. (Remember to enter each variable in its own column and each individual on its own row.)

*10. Create a new worksheet and enter the gas mileage data from problem 11.6 (pages 208 and 209) in the text. Display the distribution of this data by making a histogram and sketch its shape below (find **Histogram** under the **Graph** menu).*

Appendix D: Tables

Every statistics textbook has an appendix with a Normal Table, a t-Table, and a Chi-Square Table.

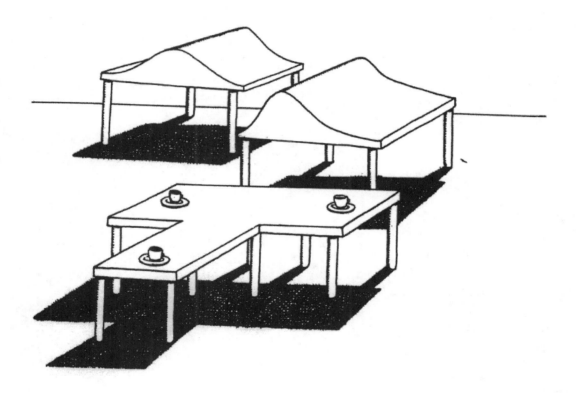

Rick Miller '92